T0205203

A NEW FRONTIER

INNOVATIONS IN COMMERCIAL SPACE TRAVEL

www.royalcollins.com

NEW
FRONTIER

INNOVATIONS IN
COMMERCIAL SPACE TRAVEL

H<small>UANG</small> Z<small>HICHENG</small>

Books Beyond Boundaries
ROYAL COLLINS

A New Frontier: Innovations in Commercial Space Travel

Huang Zhicheng
Translated by Zhang Ruofeng

First published in 2021 by Royal Collins Publishing Group Inc.
Groupe Publication Royal Collins Inc.
BKM Royalcollins Publishers Private Limited

Headquarters: 550-555 boul. René-Lévesque O Montréal (Québec) H2Z1B1 Canada
India office: 805 Hemkunt House, 8th Floor, Rajendra Place, New Delhi 110 008

Original Edition © Publishing House of Electronics Industry

ISBN: 978-1-4878-0452-7

To find out more about our publications, please visit www.royalcollins.com

CONTENTS

Chapter 9: The Opportunities and Challenges of Commercial Space Travel in China..185

CHAPTER 1

A NEW GLOBAL SPACE ERA

After the manned lunar landing in the 20th century, the global space industry reached new heights, which augured a new era that has been characterized by full commercialization as well as innovation-driven and civil-military integration. Commercial space travel has now evolved into an important aspect of the aerospace industry, along with civil and military aerospace.

1. The Development of Space Travel During the Cold War

The desire to travel to Outer Space has long been a dream for mankind over many years of understanding and transforming the world. Obviously, this was impossible for a long time due to very low productivity and the absence of science and technology. Based on remarkable advances in rockets, electronics, and automatic control, the first artificial Earth satellite was successfully launched, beginning a new era of manned space flight.

In the past 60 years, space technology has developed rapidly to become one of the most influential fields in science and technology as well as in society more widely. It brings enormous economic and social benefits, and has also become an important symbol of advanced civilization. Current space technology integrates many of the most impressive achievements in science and technology in a system of mutuality and complementarity. Space exploration has not simply been a case of mankind leaving the Earth; its impact on politics, economics, military exploits, culture, and society are so extensive that they reach beyond the bounds of scopes of science and technology.

Despite encountering many difficulties in the course of exploring space, mankind has had to persevere. The path has involved ancient rockets, the three laws of Isaac Newton (1646–1727), the multistage rocket theory of Konstantin Tsiolkovsky, and the development of the rocket by Wernher von Braun (1912–1977), as well as the theory and practice of the renowned Chinese scientist Qian Xuesen (1911–2009) at the California Institute of Technology in the field of rockets. On October 4, 1957, the Soviet Union successfully launched Sputnik 1 with the SS-6 ballistic missile. On April 12, 1961, Soviet astronaut Yuri Gagarin (1934–1968) orbited the Earth for the first time in the Vostok 1 spaceship.

The first phase of space development during the Cold War was the so-called 'space race' between the Soviet Union and the USA after World War II. In this fierce competition, the Soviet Union took the lead from the beginning.

The Soviet Union's success in space was a wound to the American psyche. In 1958, the National Aeronautics and Space Administration (NASA) was founded, and its first satellite – Explorer – was launched that year. On July 20, 1969, American astronaut Neil Armstrong (1930–2012) stepped onto the Moon from the lunar module of the Apollo 11 spacecraft. On April 19, 1971, the Soviet Union launched the first Salute 1 space station. On April 12, 1981, 20 years after Gagarin's first manned space flight, the first American space shuttle – Columbia – was successfully launched into space.

The second phase of space development during the Cold War was an era of confrontation. The United States and the Soviet Union scrambled to muster military forces to control outer space. Various military activities created a new struggle that persisted into the 21st century. Satellites, space stations, space shuttles, and other space vehicles orbiting Earth are frequently assigned to perform military tasks.

Conquering outer space has long been an unparalleled strategic high point in international competition. Six weeks after the Soviet Union launched its satellite, the US military proposed an anti-satellite program. In the early 1960s, US President Kennedy pointed out that the fight for space supremacy would be the main mission of the following 10 years, and the country that controlled space would control the Earth.

Meanwhile, US President Eisenhower did not see the USSR's successful satellite launches as a threat. In his view, the functions of satellites as weapons were very limited. He believe that the US should focus on developing reconnaissance satellites that would prove useful in the Cold War as a valuable asset. In August 1960, Eisenhower authorized the establishment of the National Reconnaissance Agency (NRO) to manage the top-secret Corona reconnaissance satellite program. Corona launched a number of experimental satellites, and began to recover photos from #14. By the time the program ended in 1972, it had carried out 145 missions, 102 of which were successful, including secretly filming all of the Soviet missile facilities. This allowed the USA to assess its own lack of missiles in comparison to the Soviet Union. During the Cuban Missile Crisis, Corona provided President Kennedy with an accurate estimate of the number of Soviet nuclear weapons. It also tracked the building, launching, and even service of new Soviet submarines, and monitored the execution status of the USSR in Stage I of the Strategic Arms Limitation Treaty (SALT I). Soon after, the USA developed its own satellites for military navigation, early warning, and communications, and transferred military satellite technology to civilian satellites.

In order to take the lead in manned space flight, the US military also took a concerted interest in the field of military manned space flight. Many military purposes were conceive for manned orbiters, including reconnaissance, anti-satellite, and ground bombing, and a series of military manned spacecraft development programs was launched, such as the 'X-20' program in the United States Air Force (USAF) and the 'Manned Orbital Laboratory (MOL) program'. The X-20 program (also known as Dyna-Soar) started in 1958, and aimed to develop a small experimental winged manned spacecraft. It was similar in shape to the later American space shuttle orbiter, and was launched with the Hercules 3 rocket. This program was canceled in October 1963 for technical and economic reasons. US military forces also conducted manned space reconnaissance tests on space shuttles in the years that followed.

The Soviet Union took a different path from the United States. In the mid-1960s, it began to design and manufacture 'Salyut' space stations. Salyut 1 was launched in April 1971, followed by seven more space stations were launched. Salyut 2, 3, and 5 were military aerospace stations for manned space travel experiments. However, for a variety of reasons, the manned space military tests between the USSR and US ceased. Two days after Yuri Gagarin went into space, Kennedy decided to launch a Moon mission in an effort to prevail in space technology. Finally, this mission was a historic success.

On March 23, 1983, US President Ronald Reagan announced the implementation of the Strategic Defense Initiative (SDI), also known as the 'Star Wars Program', which provoked an unprecedentedly fierce arms race between the US and the USSR. The plan was based on retired US lieutenant general Daniel O. Graham's 'high frontier' theory, which advocated a strategic vision for US space travel. Although this program was not the direct cause of the collapse of the USSR, it did accelerate the decline of the superpower, and was the straw that broke the camel's back.

The third phase of space development in the Cold War period was dominated by governments. Both the space race and the later confrontation were invested, organized, and managed by governments. Private enterprises were also involved due to receiving contracts for government projects in the USA and other Western countries.

With the maturing of rocket and satellite technology during this period, communication satellites became commercialized. After the collapse of the Soviet Union, aerospace development entered a period of transition. With the maturity of basic space technology, the space race and the militarization of space gave way to the comprehensive promotion of social and economic development as well as civil-military integration. The emerging 'space economy' and 'aerospace commercialization' changed all aspects of human life, beginning a new era characterized by comprehensive commercialization as well as innovation-driven and civil-military integration.

Shortly after the founding of the People's Republic of China, the Chinese government put forward a plan for the development of the aerospace industry. China's first man-made Earth satellite was launched on April 24, 1970, indicating that China had become a country with the ability to develop and launch satellites. After 60 years of work, China has made impressive achievements in space science and technology with much less investment than advanced countries. The Long March series of rockets has entered the international market with higher reliability and lower launch costs. Satellites for telecommunication, remote-sensing, navigation, and meteorology have made significant contributions to China's social, economic, scientific, technological, and military development. On October 16, 2003, astronaut Yang Liwei successfully entered orbit and returned, landing the Shenzhou V manned spacecraft in the desert. This signified China's entry into the field of manned space, which had been dominated by the USA and Russia for 42 years. China is currently making major innovations in all aspects of space science and technology, offering greater contributions to the development of society.

2.   New Missions for the Development of Space

In order to expand social production, human beings must constantly expand their fields of activity. This has involved expansion from the land to the sea, then from the sea to the sky, and finally from the sky to space. If land was the first area of human activities, the ocean is the second, the sky is the third, and space is the fourth. Each leap in the scope of activity has improved our ability to understand and utilize nature, and has promoted the development of productive forces and social progress.

As for the future development of space travel, its missions are to explore, utilize, and protect space.

The first mission of space development is to explore space. Because the Earth is only a small planet in a vast universe, our current understanding of it is limited at best. The human desire for knowledge is the eternal motive for space exploration. According to NASA, the main task of its Department of Science is to solve the fundamental problems of the Earth, the solar system, and the universe and to study their relationships with the future of mankind. The most urgent questions include how the solar system evolved, how life was formed, and are we alone in the universe. Even if it is not profitable to consider such basic questions, answering them will have a revolutionary impact on scientific discoveries.

The second mission of space development is to utilize space. The purpose of space exploration is not only to investigate the possibility of human migration to other planets, but also to protect and develop the Earth. The Earth is now facing serious threats such as climate change, ecological destruction, energy exhaustion, and asteroid strikes, which force us to consider migration. Mars is now a hot spot for human space exploration, and a likely destination for migration, although it currently seems unrealistic. Therefore, the only way for human society to achieve sustainable development is through environmental protection and sustainable use of energy. Therefore, it is an important task to develop space technology to exploit and utilize space resources in order to improve human life on Earth. Thus far, great achievements have been made in the fields of communication, navigation, and remote-sensing satellites, and the development of material resources and energy in space has begun.

Space travel has the potential to profoundly affect our lives in a variety of ways. Its development has shifted from the initial goals of entering space, exploring the universe, and safeguarding national security to a focus on the application of space technology to promote economic and social development. One of the purposes of exploring space is to make use of space resources, develop the economy, improve people's lives, and integrate space travel with public life.

At present, the proportion of the space economy in the world economy is still low, although it is increasing year by year. Therefore, there is huge potential for growth in the use of space for public benefit.

The final mission of space development is to protect space. History has proved the importance of technology. Whoever occupies the commanding heights of military technology will be able to seize advantages in times of war. History also proves that technological inventions are often closely related to military affairs. The development of advanced technology is bound to play an important role in the military field. Today, with the development of space technology, major countries have launched weapons in space. Therefore, tough diplomatic negotiations must be performed for an early agreement on the control of space armaments, creating the necessary environment for peaceful global development.

In addition, with the increase in space activities, low-Earth orbit is becoming congested, and increasing amounts of debris are beginning to affect the normal operation of spacecraft. This means that the space environment must be subject to governance.

Finally, in the process of space exploration, China must abandon the human-centered concept of conquering a new frontier. Efforts must be made to protect the ecosystem of space and achieve the ultimate goal of cohesion between nature and mankind.

3. The Limitations of Space Development

Throughout the development of space travel, technology has frequently failed to meet expectations. In order to realize the new mission of space exploration in the future, the bottlenecks must be analyzed.

First of all, the development of space technology has encountered a series of technical bottlenecks. In order to fulfill the new missions mentioned above, it is necessary to greatly enhance the capabilities of entering, exploring, utilizing, and controlling space.

In terms of the capacity of entering space, our ability to reach further destinations is obviously deficient, and there are prohibitions in terms of the costs, operation efficiency, and comfort of launch vehicles. It is therefore necessary to develop reusable technology for carrier rockets and new engines, and to conduct research into aerospace aircraft technology (of which the operation mode is similar to aircraft), as well as to explore new energy and new engine applications. The development of advanced life support systems and environmental control systems should be emphasized to improve the safety and lifesaving capabilities of manned systems. It is also necessary to deal with problems such as radiation resistance for astronauts, and effective entry, deceleration, and landing techniques for manned systems that require entry into the Martian atmosphere.

The capability, accuracy, and efficiency of space exploration are obviously inadequate. Therefore, powerful new-generation detection technology must be developed, and research should be conducted into independent technology, deep space communication technology, and new detection instrument technology. For the exploration of Mars and other planets, more efficient probe air vehicle technology should be developed in addition to improving Rover technology.

In regard to utilization capabilities, satellites for many applications such as reconnaissance, navigation, and communications must be consistently upgraded with the progress of the new technological revolution. The potential applications of micro-satellites also need to be expedited in terms of growth. The development of technologies such as space robots, laser communications, quantum communications, micro-array interference imaging, and image reconstruction will break the limitations of traditional satellite functions and have a significant impact on the development and application of future satellites. The development of the low and medium Earth orbit satellite constellation will replace existing satellite systems in terms of scope and depth, and will create the conditions for the construction of an integrated geospatial information network. The development of multi-purpose satellites and constellations (especially integrated satellites for navigation, remote-sensing, and communications) will expand the application areas of satellites. The development of cutting-edge technologies such as artificial intelligence and Big Data has opened up a broader horizon for the analysis and application of satellite data.

Further breakthroughs in space technology need to be achieved for the development of space solar energy and resource utilization for the Moon and other planets. The construction of solar power stations in space is one competitive idea for solving the global energy supply crisis. In order to realize this idea, it is necessary to reduce the cost of space transportation, investigate the possibility of complex assembly and maintenance technology for astronauts in space, and effect energy transmission through power beams and the resulting influence on the environment.

In terms of the capability of space control, new technologies such as Space Situational Awareness (SSA), spacecraft modularization, space self-assembly, and debris removal should be developed. Automatic assembly technology will greatly reduce the cost of spacecraft, revolutionize our ability to enter and use space, and display significant application value for space control. SSA is the key to understanding and controlling the intentions and trends of potential adversaries' space activities and ensuring the safety of space assets. It is also the basis of and precondition for space control. The warning time for space events will be reduced from several weeks to a few hours, and the ability of SSA will augur an exponential increase in data accuracy and affordability.

Confronted with the abovementioned needs, the development of space technology is now facing a series of tough challenges. In fact, technological development varies with time and space, and there is an imbalance between various fields of science and technology. In general, technologies such as computers and networks related to the 'bit world' have developed rapidly, while those related to the 'atomic world', such as engine power, are developing relatively slowly, especially in the field of aerospace propulsion, where there has been almost no progress for decades. Tyler Cowen's book *The Great Stagnation* discusses the slow development of modern technology, and gives examples of space technology. He writers 'The moon landing occasioned great excitement and it was heralded as the beginning of a new age. But it's more properly seen as the culmination of some older technological developments.' In fact, for space technology, the satellite field (which is related to information technology) is developing rapidly, while the space carrier field (which is related to power propulsion technology) is not.

Second, the development of space has encountered a bottleneck in terms of management. Traditional space activities are funded by the government and aim to realize the national strategy, so they are forms of national behavior. In the initial period of technological development, many countries established their own national space agencies. For example, on October 1, 1958 the US established NASA according to the Public Law 85-568 signed by President Eisenhower. Relying on the national system and concentrating on major affairs, many important space missions (including the Apollo moon landing program, the Skylab space shuttle, and the International Space Station) have been carried out in the subsequent 60 years. These national space agencies are still playing a vital role in developing their own space technology, civil space activities, and commercial space travel.

In addition, overwhelming bureaucracy and a conservative culture were fostered over the years of NASA's long-term operation. The most prominent instance was the space shuttle Challenger disaster in 1986. The previous year, Roger Boisjoly – an engineer at Morton Thiokol – and others had warned that the joints of solid rocket boosters on space shuttles could fail in cold weather, with catastrophic consequences. On the eve of the launch on January 28, 1986, Boisjoly and four other engineers spent the night trying to convince their superiors about the defects in the solid rocket boosters and the possibility of a serious accident, and recommended delaying the launch. They insisted that the O-ring seal on the rocket's joint was highly likely to fail at low temperatures, causing a leak in the high-heat and high-pressure gas that is supposed to be sealed inside the solid rocket's boosters. The gas would burn the metal shell of the rocket, leading to structural failure. However, their suggestions were rejected by the company's top management and NASA, which dismissed the delay as untenable and insisted that the shuttle be launched in sub-zero temperatures the next morning. This decision was one of the most serious engineering mistakes in history.

On February 1, 2003, the space shuttle Columbia exploded as it re-entered the atmosphere, shocking the world once again and impacting NASA's management culture. Both accidents have meant that state institutions now tend to maintain a conservative attitude in their approach to technological innovation.

With the development of space technology in the United States, a small number of enterprise groups with technical strength and experience in complicated system projects have emerged, such as Boeing and Lockheed Martin, which monopolize the space market and share government orders. This has led to a higher cost of development and launch for spacecraft. The US military's space launch has been controlled for many years by the United Launch Alliance (ULA) – a joint venture between Boeing and Lockheed Martin, which competed for military contracts until they became partners in 2006. In 2013, the Air Force awarded ULA a five-year contract worth $11 billion to produce the Evolved Expendable Launch Vehicle (EELV) for launch missions by the Atlas V and Delta IV rockets. As the only military-certified launch service provider, ULA will complete 36 rocket launch missions without competitors. It was not until 2014 that the monopoly was broken, when SpaceX's Falcon 9 rocket obtained permission to launch military satellites.

Finally, the development of the aerospace industry has encountered a bottleneck in terms of investment. In the traditional mode of aerospace, much more money was spent, and there were many constraining factors such as technology, management, efficiency, and monopoly. The Apollo Moon Landing Program cost an estimated $151 billion. The US space shuttle was launched 135 times at a cost of $196 billion, with an average cost of $1.5 billion per launch. This was too much even for the wealthy United States to afford, and the shuttle had to be retired in July 2011. In 2006, the United States proposed a 'constellation' program to return to the Moon, which was expected to cost $99 billion, and was canceled due to high cost. High investment makes it difficult to develop space travel in the traditional model.

4.   The Rise of Commercial Space Travel

The Cold War between the USA and the USSR in the 20th century was the main force in promoting the initial development of space technology. With the end of the Cold War and the gradual maturity of space technology, the space race gave way to space militarization with a view to comprehensively promoting social and economic development and civil-military integration. The emerging space economy will inevitably lead to commercial space travel.

First of all, throughout the history of space development, due to high risks and cost, more conservative technology has been employed to ensure high reliability, which hinders the pace of space innovation to a certain extent, and also makes it difficult to break through bottlenecks. After the successful completion of the Apollo Moon Landing program, the US government reduced investment in space activities to one-eighth that year. This severely limited NASA's manned space activities to low-Earth orbit for a long time.

In recent years, in order to maintain its leading position in space technology, the USA has set its sights on manned missions to Mars. On October 8, 2015, NASA published 'Journey to Mars: Pioneering Next Steps in Space Exploration', which laid out the USA's plan to carry out manned Mars exploration missions in three phases. The first phase is to integrate all resources, starting from the current 'Earth reliant' manned space program; the second step is to develop a 'proving ground' in cislunar space; and the third step is to form a capability of 'Earth independence' for deep space exploration.

On October 11, 2016, then President Barack Obama posted an article titled 'America will take the giant leap to Mars' on CNN's website, writing that 'We have set a clear goal that is vital to the next chapter of America's story in space: sending humans to Mars by the 2030s and returning them safely to Earth, with the ultimate ambition of one day remaining there for an extended time'.

After Donald Trump took office, he signed the NASA Transition Authorization Act (hereinafter referred to as the Act) on March 21, 2017, which approved NASA's budget of $19.5 billion for the fiscal year 2017. According to the Act, the overall goal of American space flight is to continue to work with its international, academic, and industry partners 'to extend human presence into the solar system and to the surface of Mars'. The Act also emphasizes that the US government will continue to encourage private companies to enter the aerospace industry. This will reduce its expenditure on space exploration, and will also make the American space industry more competitive in the global market.

On September 28, 2016, at the 67th International Astronautical Congress in Mexico, Elon Musk – the founder of SpaceX – gave a keynote speech entitled 'Making humans a multi-planetary species', and explained the 'Interplanetary Transport System' for human settlement on Mars, proposing that a manned landing on Mars could be achieved in 2025 followed by human settlement.

Secondly, the maturity of fundamental space technology has nurtured a swath of aerospace entrepreneurs. The development of commercial aerospace is driven by the dual engines of the government and the market, promoting disruptive technological innovation, and finding a low-cost, low-risk way to space. This is how commercial space travel came into being.

Commercial space travel is an independent non-governmental aerospace activity that configures resources such as technology, capital, and talent according to market rules, and aims at making a profit. It includes satellite manufacturing, launch services, satellite operation and application, manufacturing and service of ground equipment, space tourism, and the development of space resources.

The commercialization of travel began with satellite applications. With the increasing popularity of television in the 1970s, satellite and launch services stepped into the field of commercialization, following the commercialization of remote-sensing satellite and navigation satellite applications. It is now advancing into the field of manned space flight, and may even extend to some areas of space exploration. As a result, the development of the global space industry will enter a new stage of full commercialization. At this time, although national space agencies around the world still play the roles of planning, coordination, and guarantee, many private and mixed enterprises will become an important force in the development of the aerospace industry.

The US leads the world in commercial travel. In the United States, commercial space refers to the commercialization of space activities and the participation of private commercial enterprises. American commercial space projects cover three aspects: first, the commercial operation of national space projects means that major national space projects (including military operations) are contracted to private enterprises through the market competition mechanism; second, the commercial operation of non-governmental space projects means that space projects are initiated by private enterprises, funded by society, and operated independently at their own risks. Third, the commercialization of space applications means that private enterprises provide products and services in the fields of satellite communications, remote-sensing satellites, and satellite navigation.

A large number of start-ups have emerged in the United States in various fields such as space transportation systems, satellites and their applications, space stations, space tourism, deep space exploration, and the development of space resources. Aerospace has become one of the most aggressive and promising entrepreneurship fields in the United States after the Internet. Development is also moving from being government-led to the joint development of national teams and commercial aerospace companies.

In Europe, commercial space travel refers to all space activities that operate in accordance to market rules. In Europe, unlike the United States, many mixed enterprises are involved in commercial space activities.

In addition to the basic characteristics of high technology, high investment, and high risk, the commercial space industry has the following characteristics.

First, it is contemporary. Commercial space travel is the result of the development of space technology to a certain stage, and the progress of the aerospace industry on a certain scale. Commercial space travel was formally proposed by the USA towards the latter stages of the Cold War. In 1982, the US National Space Policy divided space activities into military and public sectors, while military aerospace activities were classified into two categories: confidential and non-confidential. In 1988, according to the requirements of space development, the US National Space Policy formally separated commercial space from military and public aerospace.

Second, the commercial space travel industry is profitable. Unlike military and public space travel, it is not directly planned by the government, and mostly receives investment from private enterprises. Pursuing benefits and maximizing profits is therefore an important feature of commercial space flight.

Third, the commercial space industry is driven by the market. Commercial space activities are market-driven, and the development of business operations are in accordance with the market mechanism and market rules. The government should also purchase commercial aerospace products or services according to the principle of open competition.

Fourth, the commercial space industry is restricted. Although commercial aerospace operates in the way of the market mechanism, it is not a completely liberal economy. It must also be restricted by international rules and national laws. Its development should guarantee national and public security. For example, America's commercial space cooperation with foreign countries must comply with the US national export control system for space technology products.

Foreign cases show that commercial aerospace has obvious advantages over traditional aerospace in terms of technological innovation, efficiency improvement, and cost reduction. It also plays a significant role in promoting economic development. The development of commercial aerospace can enhance the benefits of national investment, and promote civil-military integration in the aerospace industry.

China's aerospace development should draw from the successes of other countries in developing commercial aerospace, and take effective measures to promote its development, in order to advance innovation and international competitiveness and transform China from merely a large player in global aerospace to a strong one as well.

5. Entering the New Space Age

The global space industry reached a new height after the manned lunar landing in the 20th century, which augured a new era that was fully commercialized and innovation-driven, with civil-military integration. Commercial space travel has now evolved into an important aspect of the aerospace industry along with civil and military aerospace.

The first feature of the new space age is full commercialization. Commercialization is an inevitable result of the development of the aerospace industry, and will become a new driving force of unprecedented prosperity and development in the aerospace industry. The commercialization of space travel will inevitably lead to commercial space flight.

The United States began to encourage the development of commercial space travel in the early years of the George W. Bush administration. Barack Obama also showed a special preference for this sector. In the spring of 2010, he visited the SpaceX launch site to talk with Elon Musk's team just before he officially announced that Bush's 'Return to the Moon' program would be replaced by a manned Mars program.

On June 28, 2010, the US government announced a new National Space Policy, vowing to commit itself to commercializing space activities in the future, providing more opportunities for the business community to participate, and promising that the relationship between the government and the business sector would be cooperative, not competitive. At the same time, it proposed various incentive mechanisms to encourage the commercialization of space activities.

NASA believes that commercial companies in a free competitive market can better afford to develop and operate space programs than government agencies. In November 2005, Michael Griffin, then director of NASA, stated: 'I believe that with the advent of the ISS (International Space Station), there will exist for the first time a strong, identifiable marked for "routine" transportation service to and from the LEO (Low-Earth Orbiter).' He went on, 'especially after we retire the shuttle in 2010, but the capability should become available earlier. We want to be able to buy these services from the American industry to the fullest extent possible.' At the same time, the impact of the financial crisis tightened NASA's budget, and intensified the need for cost reduction.

By reducing investment and improving efficiency, the US government hopes to free itself from traditional space projects with relatively mature technologies and low risks through commercial space travel, so as to focus on the development of cutting-edge space technologies and manned deep space exploration. In order to promote commercial space flight with crews and cargo transportation, NASA proposed the 'Commercial Transportation Plan', encouraging private companies to develop a commercial orbit transportation system. Through commercial projects, NASA hopes to transform its partnership with private companies from buying spacecraft to buying services. Commercial companies are responsible for the entire operation of the project, including system design, research and development, manufacturing, testing, launching, and operation management. NASA is only responsible for the schedule, safety supervision, and technical support, as well as the corresponding start-up financial support for commercial aerospace companies. This model will give commercial space companies greater autonomy and flexibility.

The second feature of the new space age is that it is innovation-driven. The continuous innovation, integration, and application of technology have become a new driving force for development of commercial space travel.

Within the new scientific and technological revolution and industrial transformation around the world, the Internet is profoundly affecting global space development, especially commercial aerospace. In recent years, the cross-border integration of aerospace and the Internet has become a trend. The Internet has a disruptive impact on aerospace development by means of Internet thinking on commercial aerospace, providing application services via online platforms, exploring the creation of a 'space Internet', and moving Internet companies into the aerospace market.

Looking to the future, low-cost and reusable launch technology, advanced propulsion technology, artificial intelligence, robot technology, Big Data technology, micro system technology, 3D printing, new energy, and new materials will be widely used in the space field, thus stimulating the development and transformation of the commercial space industry through research, launching, operation services, and satellite application.

Although the business model of commercial space is important, the key to breaking the bottleneck of traditional space travel is to be able to select disruptive technology projects in the space field to achieve effective innovation.

The third feature of the new space age is civil-military integration. In order to meet the growing demand for military aerospace, the US military has steadily increased the use of commercial space resources and services in many fields, such as entering, using, and even controlling space. With the acceleration of commercialization, commercial space resources and services will continue to infiltrate the military aerospace field, providing important support for the development of related equipment and technology. Commercial aerospace has become a breakthrough in the in-depth development of civil-military integration.

In terms of entering space, American commercial companies have undertaken official launch missions into military aerospace. Space exploration technologies have been approved for the launch of US military satellites, such as reconnaissance satellites, third-generation GPS navigation satellites, and the US Air Force's small X-37B unmanned space shuttle.

In the use of space, commercial satellites have become an important source of space-based information support for the US military. Commercial remote-sensing satellites have provided a large number of high-resolution target images in support of the Libyan war, and have monitored missile tests and nuclear facilities in North Korea and Iran. This has provided important support for the US military to improve the efficiency of fixed-point strikes, and obtaining information for diplomatic initiatives. Around 20% of the total data flow in the Gulf War and 80% in the Iraq War were provided by commercial communication satellites for the US military satellite communications. The US is now carrying out payload tests on commercial satellites for missile early-warning, and will carry a 10 kg payload for damage assessment on the next-generation Iridium commercial satellite constellation in order to evaluate the effect of missile-defense interceptor missiles.

In the field of controlling space, commercial companies are providing situational awareness services for the military and government departments using software for Big Data management, compensating for the inadequacy of the US Air Force in its situational awareness capabilities.

In the field of space protection, the use of commercial satellites to carry military loads has become an important way of implementing the flexible and decentralized space system architecture in the United States, which will enhance the security and anti-destruction capacity of the US military aerospace system. In addition, the United States Strategic Command has begun to increase the number of commercial space companies in the Joint Space Warfare Center, strengthen cooperation between the military and the commercial space industry, and schedule and coordinate the use of commercial satellite resources to serve space military operations in a timely manner.

References

[1] Qian Xuesen. *An Introduction to Interstellar Navigation* [M]. Beijing: Science Press, 1963.

[2] Huang Zhicheng. *Sky and Sky Vision* [M]. Beijing: Electronic Industry Press, 2015.

[3] Huang Zhicheng. *The Fourth Wave of Aerospace Science, Technology, and Society* [M]. Guangzhou: Guangdong Education Press, 2007.

[4] Huang Zhicheng. 'Commercial Aerospace Leads the Aerospace 2.0 Era' [J]. *International Space*, 2017 (3): 2–6.

[5] Huang Zhicheng. *Commercial space travel begins a new era* [N]. Learning Times, 2017–04–24 (3).

[6] An Hui. 'Commercial space travel: new vitality for the US aerospace industry' [J]. *Space Exploration*, 2016 (4): 26–31.

[7] Zhang Zhenhua, Bai Mingsheng, Shi Yong, et al. 'The Development of Foreign Commercial Aerospace' [J]. *China Aerospace*, 2015 (11): 31–39.

[8] Wu Qin, Zhang Mengtian. 'Analysis of the development of US commercial aerospace' [J]. *International Space*, 2016 (5): 6–11.

[9] Huang Zhicheng. 'How space science will affect the future of mankind' [J]. *Space Exploration*, 2017 (6): 1.

[10] Huang Zhicheng. 'Commercial space travel: a new driving force for space exploration in the United States [J]. *International Space*, 2016 (5): 64–68.

[11] Tyler Cowen. *The Great Stagnation* [M]. Tu Zipei, Trans. Shanghai: Shanghai People's Publishing House, 2015.

CHAPTER 2

CULTIVATING DISRUPTIVE SPACE TECHNOLOGY

Decisive opportunities for innovation in science and technology depend on breakthroughs in disruptive technology. The developed countries of the world have made disruptive innovation an important way of enhancing their abilities in innovation for science and technology. Due to the high-risk and high-input characteristics of the aerospace industry, the development of disruptive space technology is facing severe challenges. However, only by realizing disruptive technological innovation in space can China break through the bottleneck in current space development in order to go faster, deeper, and better.

1. Cultivating Disruptive Space Technology

Clayton Christensen, a professor at Harvard Business School, and his colleagues first proposed the concept of 'disruptive technology' in a paper titled 'Disruptive Technology – Catching the Wave' in 1995. It is a kind of technology that creates a new path and has a disruptive effect on the existing traditional or mainstream technological approaches. It may be an all-new innovation technology, or an interdisciplinary and cross-field innovative application based on existing technology. Disruptive technology breaks the thinking and development route of traditional technology, and represents a leap forward in the development of traditional technology. For instance, digital technology has been applied to the field of photography, disrupting the traditional film-based technology. The computer network technology produced by the fusion of computer technology and communication technology has overturned the traditional mode of information transmission and application.

Christensen published a book called *The Innovator's Dilemma* in 1997, in which he first proposed two kinds of innovation: sustaining and disruptive. Sustaining innovation is a step-by-step advancement for enterprises in improving on existing technologies and products. Disruptive innovation creates new technologies or applications and provides customers with products that are simpler, more convenient, and cheaper. In fact, when technology and markets change dramatically, many exceptional enterprises – which had been regarded as models or examples for imitation with their traditional technology and non-disruptive innovations – not only lost their leading position in the industry, but also left the market completely, such as Kodak, Nokia, and Motorola.

1.1 The characteristics and significance of disruptive innovation

The United States refers to emerging technology (i.e. forward-looking, pioneering, and exploratory) as 'frontier technology' and more recently 'breakthrough technology'. These terms have the same meaning as 'disruptive technology', which was introduced from the academic community in China. All of these phrases emphasize the revolutionary nature of technology, and the fact that it changes the rules of the game. 'Breakthrough technology' emphasizes the development of the technology itself – changing at the beginning, then developing rapidly along a new technological path but when it reaches a critical point (singularity). As for the widespread use of the term 'disruptive technology' in China, it does not distinguish between 'breakthrough technology' and 'disruptive technology'. However, this book still uses the term.

Looking back at the development process of disruptive innovation in developed countries, especially the USA, disruptive innovation has the following characteristics:

> First, it is a 'revolutionary' and 'destructive' form of innovation, breaking the traditional rules of the game. It inevitably encounter resistance, so there will be a process of succession from the new to the old. Philosophically, the most essential implication of innovation is 'destructive construction', and there is a similar understanding within economics. Joseph Alois Schumpeter, an economist known as the founder of innovation theory, wrote: 'The new combination meant the competitive elimination of the old combination'. In the past, the saying went that we see further because we stand on the shoulders of giants and follow the path of our predecessors. In the era of innovation, this should probably be changed: we can see further because the giants have been knocked down to the ground; and only when we start a new path can we go further. At the same time, disruptive technology is a double-edged sword that brings many risks as well as advantages. Whether it has a negative impact on social development and whether it involves ethical issues will also be debated for a long time to come, thus increasing the complexity of disruptive technological innovation.

Second, disruptive innovation is 'whole chain' innovation. In fact, shortly after Christensen put forward the concept of disruptive technology, he replaced the concept with the idea of 'disruptive innovation', aiming to emphasize the 'disruptive effect' brought about by technology rather than the technology itself, and complete a 'whole chain' transformation process from technology to engineering. This process usually takes many decades. The US Defense Advanced Research Projects Agency (DARPA) invented an early form of the internet, known as ARPANET, in 1967, and it was only in 1991 that the Internet as we know it emerged. In the process of technological innovation, DARPA is only responsible for the demonstration and verification of technology. For example, in the 1970s, it developed Have Blue, a proof of concept project for stealth aircraft. Later, technology was transformed into engineering and practical applications. For example, the US Air Force was responsible for the development of F-117 stealth aircraft. In the process of transforming technology to engineering and its practical application, many DARPA-developed technologies have been aborted or eliminated. For example, the F-117 stealth aircraft was decommissioned shortly after passing the test for combat in Kosovo due to poor performance. However, its technology still laid the foundation for the development of a new generation of stealth aircraft in the United States.

Third, there are many uncertainties around disruptive innovation. According to Qian Xuesen, modern science and technology is an open and complex giant system, in which every subsystem will change over time, and for which development is unbalanced. It is difficult to predict what sort of disruptive technology will emerge, and when. This is very similar to the Earthquake phenomenon. Although the occurrence of Earthquakes is governed by natural law, it is still very difficult to correctly predict when and where they will occur. Disruptive technological innovations are similar. For instance, scientists in the 1960s did not expect LSI and computer miniaturization to lead the rapid development of information technology and the Internet. So, disruptive innovation is not 'planned'; it can only be anticipated with policies and measures.

We are now in the eve of a new technological revolution, which will lead to a new industrial revolution. The first industrial revolution used steam as the power to mechanize production. The second achieved mass production through electricity, while the third used electronics and information technology to achieve automation of production. The fourth industrial revolution will be characterized by the integration of various technologies, and will eliminate the boundaries between the physical, digital, and biological worlds. This revolution is predicted to radically change the way we live, work, and socialize. We do not yet know how it will unfold, but in terms of its scale, scope of influence, and complexity, it will be completely different to anything mankind has experienced before.

Disruptive technology is the leader of this new industrial revolution. With their unprecedented capabilities in processing and storage, mobile devices are an easy source of knowledge, connecting billions of people and releasing unlimited innovation potential. At the same time, the latest breakthrough innovations and technologies (such as AI, robots, unmanned systems, Big Data, the IoT, 3D printing, Nanotechnology, Biotechnology, materials science, energy storage, and quantum technology) are creating infinite possibilities. The fourth industrial revolution is disrupting almost all industries in all countries. These changes will have an extremely wide and far-reaching impact, completely changing the entire production, management, and governance system.

The industrial revolution in the military field is an evolution in the form of war. The second industrial revolution gave birth to mechanized war, while the third sparked the information war. The fourth is set to bring forth intelligent warfare. A combat system based on information and networks will greatly improve operational efficiency, but will also mean hidden dangers and risks for its own security. With the consistent development of information warfare, the cost of weapons and equipment is getting higher. The proportion of information equipment in a weapon system is increasing, and electronic attacks and interferences are becoming increasingly serious. With the development of information weapon systems, the logic of war tells us that the efficiency-cost ratio, autonomous combat capability, and reliability of weapon systems will be improved through intellectualization.

1.2 Predictions for disruptive innovation

Obviously, whether an emerging technology can bring about disruptive innovation needs to be tested by its after-effects. Prediction is therefore necessary, but it comes with a lot of uncertainty. In the early stages of developing a new technology, it will attract the attention of the government, the media, and the investment community through extensive publicity. Expectations will rise, greatly exceeding its possible effect.

In 2008, Jackie Fenn and Mark Raskino from Gartner – an American IT consulting firm – published a book called *Mastering the Hype Cycle: How to Choose the Right Innovation at the Right Time*, which was translated into Chinese by CITIC Press as *Precision Innovation*. The book offers examples of how various technological processes have been predicted and inferred since 1995, and proposes a 'hype curve' in which the vertical axis is the display value of new technology (Visibility) and its horizontal axis is time (Maturity). From there, it divides the development of emerging technologies into five stages: the 'Technology Trigger', the 'Peak of Inflated Expectations', the 'Trough of Disillusionment', the 'Slope of Enlightenment', and the 'Plateau of Productivity'.

This model is composed of a time-varying curve of the emerging technology both in Visibility and Maturity. Such curves will vary in different fields and at different times. By comparing the position changes of the same technology on the curve, we can judge the maturity of these technologies and estimate the time when they will reach maturity. To avoid using the word 'hype', the Chinese version translates 'hype curve' into 'technology maturity curve'. In fact, the authors of the book believe that 'hype is everywhere' and proposes that 'we are left with a spectrum of behavior that involves overstating the case for an innovation to varying degrees in order to attract attention'. The authors also argue that the expected expansion of emerging technologies is due to a desire for new things, and the so-called 'conformity effect'. Of course, behind excessive 'hype', it is easy to find evidence of its origin.

The development of new technology always encounters difficulties. High expectations of practical application are difficult to meet, and often cause an anticlimax. If an emerging technology can continue to improve and mature, it will eventually be widely used. However, many emerging technologies have fallen into the 'Valley of Death' after an initial bubble of hope, a lack of funds, or even defects of the technology itself.

For any emerging technology, it is necessary to identify which stage of the curve it is on. In the 'Peak of Inflated Expectations', innovators should think calmly. In the 'Trough of Disillusionment', they need to regain confidence. Obviously, only the technology that can get out of the 'Valley of Death' can become disruptive. However, not every mature emerging technology is disruptive, and the level and influence of these new technologies should be monitored and evaluated more deeply. Developed in the USA, the Horizon Scanning method can be used as a reference.

In fact, in the process of technological innovation to date, although there are a few disruptive innovations, most are gradual. Even with these disruptive technologies, there is a long evolutionary process in which the impact extends from local to global. Therefore, China must pay equal attention to these two innovation models.

2. Disruptive Technological Innovation in the USA

The innovation ability of a country, especially of a disruptive nature, is closely related to the environment for education and innovation. Government policies play a significant role in promoting disruptive innovation, and the USA is worth learning from in this regard. the US Defense Advanced Research Projects Agency (DARPA), Lockheed

Martin's Skunk Works, the US Small Business Administration (SBA) and Silicon Valley have all provided plenty of cases. NASA's experience in cultivating disruptive aerospace technology will be discussed later.

2.1 DARPA's innovative success

For more than half a century, DARPA has been the world's most renowned leader in disruptive technological innovation. Almost all modern weapon systems in the USA military are inseparable from the strategic frontier technology developed by DARPA. As a result, several developed countries around the world are trying to build their own versions. Meanwhile, the international academic community has also begun to explore DARPA's innovative model.

DARPA was established in February 1958 by then President Eisenhower in response to the USSR's leading position in space technology. It is committed to organizing high-risk strategic frontiers for 'game-changing' projects with smaller investments to maintain the USA's global technological advantage. DARPA has nurtured and promoted major disruptive technologies such as the Internet, stealth technology, the global positioning system (GPS), lasers, and unmanned systems. It has also provided a technical foundation for the USA to enhance its military strength, and has become a booster for economic and social development. These strategic frontier technologies have played an important role in ensuring the leading position of the US military and enhancing its comprehensive national strength. The USA has since set up a number of similar agencies specializing in the development of disruptive technologies in other government departments.

In the preface to the Chinese translation of American science writer Michael Belfiore's book *The Department of Mad Scientists: How DARPA Is Remaking Our World, from the Internet to Artificial Limbs*, China's Minister of Science and Technology Wan Gang writes, 'We can see how DARPA's basic research and cutting-edge exploration in science and technology have profoundly affected the future development of the USA and the rest of the world.'

Why does this institution, which has limited funds and only about 200 staff, exert such a strong influence on US defense and world progress in science and technology? First, forward-looking innovation is encouraged. Tony Tether, who was the director of DARPA between 2001 and 2009, believes that ideas are everything. 'The best DARPA program managers, I swear, are science fiction writers,' he once remarked. DARPA encourages sharp-eyed 'mad scientists' to explore demands and identify individuals and units suitable for projects. For example, its recently launched Falcon program supports Elon Musk's SpaceX, and has successfully developed the Falcon 1 rocket. Throughout its history, DARPA's program managers have prided themselves on remaining open-minded about ideas that might at first glance appear outlandish or even absurd. Tether has said, 'Over the years we have found that the best way to prevent technological surprise is create it.'

Secondly, the tenure of program managers is what keeps DARPA alive. It employs new program managers from the industry, universities, laboratories, and the military every four years, which means that an average of 25% of program managers are replaced each year. Project management and promotion are completely irrelevant. According to Tether, 'When someone walks in the door here, there's a sense of urgency that you don't find anywhere else in government.'

Finally, there is a tolerance for failure, as DARPA emphasizes 'high-risk, high-payoff'. It does not avoid risks, preferring to manage them. Tether states, 'The projects that we do are typically projects where the idea could end up with great capability, but there's extraordinarily little data to prove that these ideas could be developed. DARPA's role and uniqueness are that we will take a bet on an idea, where other people will not.' Anyone familiar with DARPA knows that its total number of failed, half terminated, or highly variable projects will not be less than the number of successful projects. For example, in the mid-1980s, it proposed the National Aerospaceplane Program (NASP), but it was later canceled because the technology was too advanced. DARPA was not discouraged. It immediately began to explore new aerospace technologies.

However, most of the previous DARPA directors do not think that the agency has a fixed model, seeing it as using different methods for different projects at different times. On the occasion of the 50th anniversary of its establishment in 2008, DARPA interviewed 15 of its successive directors. In the interview, many of them stated that the organization's disruptive innovation should be attributed to 'serendipity'. Tony Tether said, 'I firmly believe that strategy is always best written after the fact. After the fact, I can come up with a great strategy of how we got there. But a lot of it is just pure serendipity. And that really was the strategy, to make sure that we always turn over enough rocks so that someday one of them will have a diamond underneath it.'

Is DARPA's success truly down to 'serendipity'? Of course not. There is also a unique management mechanism and innovation culture behind it. Rather than the modest words of DARPA directors, 'serendipity' or more precisely 'proper assessment' is an important factor in all disruptive technological innovations. It is related to the characteristics of disruptive technological innovation itself, as well as the general laws of science and technology. As a result, DARPA should always put forward new ways of developing disruptive technology according to changes in the international situation and the progress of science and technology.

2.2 The 14 Skunk Works principles

Skunk Works is the pseudonym of Lockheed Martin's Advanced Development Program (ADF), which is responsible for a number of aircraft designs including the U-2 reconnaissance aircraft, the SR-71 Blackbird reconnaissance aircraft, the F-117 Nighthawk fighter, and the F-35 Lightning II fighter. Skunk Works is among the top military secrets in the United States.

Although Skunk Works' projects are confidential, its methods have become some of Lockheed Martin's most valuable assets, and have started to spread across the United States. So what is their secret? The answer is simple. First, it employs as few people as possible, and each member of the team must have a strong technical ability. This team must have a 'soul character', namely its founder and renowned aircraft designer Kelly Johnson.

In general, more people mean more power, but in Johnson's opinion, more people will only reduce the efficiency of execution and lead to overstaffing. When Lockheed built the Agena-D multi-purpose satellite for the US Air Force, the department in charge of quality control for the project had a very large team of 1206 engineers. Lockheed then set up a Skunk Works group for the Agena-D program, cut out all unrelated personnel, and decentralized responsibility. At the same time, all of the red tape and bureaucracy were cut, drawing programs were changed, and design papers were sent to the workshop for production immediately instead of a month later as before. As a result, a total of 50 million US dollars was saved, the development time was cut in half, and the reliability of the satellite was increased to 96.2 percent.

Johnson outlined 14 principles for Skunk Works, which were promoted quickly throughout the United States and became unified standards for the development of the US military. A Skunk Works manager must be delegated to exert complete control of his/her program in all aspects. He/she should report to a division president or higher. Strong but small project offices must be provided both by the military and industry. The number of people connected with the project must be stringently restricted, using a small number of good people (10% to 25% less than so-called normal systems). A very simple drawing and drawing release system with great flexibility for making changes must be provided. There must be a minimum number of reports required, but important work must be recorded thoroughly. There must be a monthly cost review covering not only what has been spent and committed, but also projected costs to the conclusion of the program. A contractor must be delegated, and must assume more than the normal responsibility to secure good vendor bids for subcontracts on the project. Commercial bid procedures are very often better than military ones. The inspection system currently used by the Skunk Works, which has been approved by both the Air Force and Navy, meets the intent of existing military requirements and should be used on new projects. Basic inspection responsibilities should be pushed back to subcontractors and vendors, and inspections must not be duplicates. The contractor must be given the authority to test his/her final product in flight. He/she can and must test it in the initial stages. If he/she doesn't, he/she rapidly loses his/her competency to design other vehicles. The specifications that apply to hardware must be agreed well in advance of contracting. The Skunk Works practice of having a specification section stating clearly which important military specification items will not knowingly be complied with, and therefore is highly recommended. The funding of a program must be timely so that the contractor does not have to keep returning to the bank to support government projects.

There must be mutual trust between a military project organization and a contractor, involving very close cooperation and liaison on a day-to-day basis. This cuts down misunderstandings and correspondence to an absolute minimum. Access by outsiders to the project and its personnel must be strictly controlled by appropriate security measures. Because only a few personnel will be used in engineering and most other areas, ways must be provided to reward good performance by pay, not based on the number of personnel supervised.

Skunk Works is a common method in the aviation industry. It uses a small number of high-level talents to concentrate on breaking through key technologies and developing new models. It has now been extended from Lockheed to the outside world, and has made outstanding achievements. Its secret is to solve the most complex engineering problems with the fewest people by the simplest and most direct methods. The prosperity of the American aviation industry lies not only in the fact that it has produced so many classical aircraft, but also in the far-reaching impact of innovation on the whole industrial management.

2.3 SBA and innovation for small business

The main force of disruptive technological innovation in the United States is enterprises, especially small business that are starting up (small enterprises in the United States are equivalent to small and medium-sized enterprises in China). The driving force of enterprise innovation comes both from market demand, and from the pressure of market competition. Competition leads to the survival of the fittest. In order to ensure that they are not eliminated in the competition and can gain competitive advantages, enterprises must enhance their competitiveness through a variety of means. With the development of science and technology, product competition is increasingly characterized by technological innovation, particularly of the disruptive variety. Mastering the results of disruptive innovations means taking absolute advantage of the market. At the same time, the opening of the market often determines what kind of innovation effect this competition can bring. Only by fully opening the market promote disruptive technological innovation be promoted. For example, before 2006, the field of commercial space travel in the United States was occupied by only two giant companies, Boeing and Lockheed Martin. As soon as the decision to open the market was announced, a number of start-up companies such as SpaceX, ATK orbital, and Sierra Nevada Corporation (SNC) emerged, nurturing technological breakthroughs in the field.

Compared with large enterprises, small businesses do not have the advantages of capital, talent, and equipment. However, they are small in scale, flexible in operation, highly specialized, agile in technological innovation, and high in efficiency. They are also simple in terms of management level, strong in cohesive force, and rich in cooperative spirit. They also have the potential for disruptive technological innovation

Small businesses in the United States have shown a high level of quantity, quality, and efficiency in the cycle of technological innovation. Research shows that the number of technological innovation achievements and new technologies created by small enterprises in the United States accounts for more than 55% of the national total, and there are two technological innovations per person in small firms – five times more than in large ones. According to a survey by the Small Business Administration (SBA) established in 1958, all of the 65 inventions and creations that made significant contributions to the United States and the world in the 20th century were created by small businesses, including airplanes, computers, copy machines, and ballpoint pens.

The SBA has provided a mature approach that is driven by the dual engines of the government and the market. A comprehensive and professional service institution, the SBA was established by the Federal Government of the United States to provide financial support, technical assistance, government procurement, emergency relief, and market development (focused on international market). The SBA claims that small businesses are the cradle of innovation thanks to their rapid action, flexibility, and entrepreneurial spirit, and that the whole of society can benefit from them. The SBA's policy is to stimulate technological innovation in small businesses in the United States driven by the dual engines of the government and the market. The Small Business Innovation Research Program (SBIR) and the Small Business Technology Transfer Program (STTR) have been launched with the support of well-known international enterprises such as Microsoft and Intel. The USA's leading position in cutting-edge technology has become a successful model for the rest of the world.

The highly competitive SBIR program encourages domestic US small businesses to engage in Federal Research/Research and Development (R/R&D) that has the potential for commercialization. Through a competitive awards-based program, the SBIR enables small businesses to explore their technological potential and provides the incentive to profit from its commercialization. By including qualified small businesses in the nation's R&D arena, high-tech innovation is stimulated and the United States gains entrepreneurial clout as it meets its specific research and development needs. According to the Act, each year, Federal agencies with extramural research and development (R&D) budgets that exceed $100 million are required to allocate 2.5% of such budgets to these programs, increasing by 0.1% annually after 2011 and no less than 3.2% by 2017. Currently, 11 Federal agencies participate in the SBIR program including the Department of Agriculture, the Department of Commerce, the Department of Defense, the Department of Education, the Department of Energy, and NASA.

Modeled on the SBIR program, the STTR was established as a pilot program by the Small Business Technology Transfer Act of 1992 (Public Law 102–564, Title II). Central to the program is the expansion of the public/private sector partnership. The unique feature of the STTR program is the requirement for small businesses to formally collaborate with a research institutions in order to bridge the gap between the performance of basic science and the commercialization of the resulting innovations. The mission of the SBIR program is to support scientific excellence and technological innovation through the investment of Federal research funds in critical American priorities in order to build a strong national economy. The programs' goals are to stimulate technological innovation, foster technology transfer through cooperative R&D between small businesses and research institutions, and increase the private sector commercialization of innovations derived from federal R&D. Each year, Federal agencies with extramural R&D budgets that exceed $1 billion are required to reserve 0. 3% of such budgets for STTR awards to small businesses, increasing by 0.05% every two years after 2011, and no less than 0.45% by 2016. Currently, five agencies are participating in the STTR program, including the Department of Defense, the Department of Energy, the Department of Health and Human Services, and the National Aeronautics and Space Administration. In NASA's 2017 budget, it will increase spending on the SBIR and STTR programs from 197 million US dollars to 213 million in the 2015 fiscal year.

As a unified macro management department, the has established related policies and procedures, and has also supervised and analyzed the participating Federal agencies, promoting the sound development of the whole management system. In addition, the participation of Federal agencies involved in various technical fields improves the specialization, advancement, and feasibility of the project. Multi-party participation fully reflects the needs of the government and the market. Different Federal agencies put forward project guidelines according to their own policy objectives and market needs. After several phases, including the evaluation of commercialization potential, the project comes closer to fulfilling the needs of the government and the market. The phased support process helps to improve the efficiency of the allocation of funds. Phase I focuses on supporting a larger number of small businesses, and the amount of funds is relatively small. Phase II is based on the implementation of Phase I, which greatly reduces the risk of the results of transition and provides a guarantee for large funds for key projects. Phase III pursues commercialization for the results of Phases I and II, and improves the competitiveness and standard operation level of the projects through the market-oriented mechanism.

2.4 The new Silicon Valley culture

With the support of the SBA and the venture capital mechanism, a number of high-tech parks sprang up in the United States in the 1980s, including Silicon Valley. Also known as Santa Clara Valley in northern California, it is home to Stanford University as well as many high-tech companies and venture capital firms, and has become an important base for cultivating disruptive technology in the United States. These parks were responsible for the creation of large-scale integrated circuits and computer miniaturization, leading to the last industrial revolution. They have also spawned a new generation of disruptive technologies, such as smart phones, smart robots, Cloud computing, Big Data, 3D printing, the Internet of Things, shale gas development, electric cars, and small satellite constellations.

As the pioneer of the IT industry in the United States and the world more widely, Silicon Valley first rose to prominence for the research and production of semiconductor chips. It is now home to more than 1 million technical staff, and has an annual output value of more than 700 billion US dollars. It has bred a large number of well-known high-tech companies including Apple, Google, Intel, HP, and Cisco, and has formed industrial clusters for microelectronics, information technology, new energy, and biomedicine.

The industrial culture of Silicon Valley is at the heart of high-tech industry and the development of venture capital. Its essence is entrepreneurship and innovation. The openness of production structure in Silicon Valley is highly conducive to the flow of talent and resources. The companies there are highly tolerant of entrepreneurial failure, regarding it as a valuable asset in stimulating innovation among employees. Silicon Valley is a diversified society, with people of many races and cultures. Talent and specialty are the keys to determining personal position.

The development and growth of firms in this area depends on venture capital (VC). Silicon Valley is the center of VC in the USA. One third of all VC companies (around 200 firms) are located in the vicinity of Stanford University. Start-ups can take the critical step from technology to the market with the help of VC. As an exceptional incubator of high-tech enterprises, VC companies offer advice start-ups about financial and technological development, and promote the rapid upgrading of technology.

Instead of vertical alliances, most firms in Silicon Valley carry out on-demand procurement from networks of suppliers, creating a network of companies that are very flexible in terms of merging and restructuring. Most firms in Silicon Valley are small in scale, and find it impossible to meet all of the production requirements only by their own capabilities. This has resulted in high outsourcing needs. The powerful outsourcing support systems in Silicon Valley can turn concepts and ideas into products for small-scale industrial production. These companies work together to create the Silicon Valley ecosystem. In addition. Information exchange in Silicon Valley is very fast, and there are various channels of information exchange, allowing for efficient information exchange.

3. The Development of Disruptive Space Technology

3.1 The Characteristics of Disruptive Space Technology

In fact, disruptive technology in space is far more complex than Professor Christensen has described. The author considers that it has the following characteristics:

1) The high risk of space flight means that human space activities are complex and arduous, and the future is full of challenges. Due to high risks and high cost, space missions generally use mature and conservative technologies to ensure high reliability. The application of new technology requires a balance between advancement, reliability, and safety.

2) Whether an emerging technology can produce disruptive innovations in space travel needs to be proved with real results by means of flight verification, so the development cycle of new aerospace technology will take a long time to mature.

3) Unlike general civilian products, space missions have both government tasks and market demands, with different technologies for different tasks. Therefore, the development of disruptive space technology requires the dual drivers of the government and the market.

4) Aerospace is a form of systems engineering. At the various levels, the demands for disruptive innovation are different. Developing new aerospace engineering systems is more complicated than inventing a new product, as it not only relies on breakthroughs in basic science, but also integrates a small amount of disruptive technology into a large number of progressive technologies. This integration must be subordinated to the overall objectives of the overall space engineering system. Sometimes disruptive technologies that appear to be locally superior may still be

abandoned because they do not meet global requirements. Only in this way can China achieve 'game changing' innovation within space system engineering.

Based on the above analysis, German space experts have suggested that innovation in the aerospace field is based on a pyramid structure. The bottom layer is composed of breakthrough discoveries such as graphene, Big Data technology, quantum communication, and quantum computing. The middle layer is consisted of disruptive aerospace technologies such as space 3D printing, all-electric propulsion satellites, and nuclear rocket engines. The top layer is made up of game-changing innovations such as near-space hypersonic vehicles, space solar power stations, and space elevators.

3.2 Disruptive Space Technologies

The following are some of the disruptive innovations that NASA is developing for future missions.

1) NASA is working with the industry and the US Air Force Space and Missile Systems Center (SMC) to launch the Green Propellant Injection Task (GPIM). High-performance non-toxic green propellant is a hydroxylamine based fuel/oxidant mixture called AF-M315E. The GPIM project will conduct a flight test using the propellant to replace the current general hypertoxic hydrazine and composite binary propellant systems. This will greatly enhance the performance and volume of spacecraft.

2) Solar Electric Propulsion technology (SEP) technology will use large scale solar cell arrays to convert collected solar energy into electrical energy and deliver it to high-efficiency and energy-saving propellers to provide uninterrupted thrust. The SEP thruster uses 10 times less propellant than conventional chemical propulsion systems. Ground tests on large-scale high-power foldable solar cell arrays have been completed.

3) The Deep Space Atomic Clock (DSAC) program has completed the integration and ground testing of a small sized, highly accurate mercury ion atomic clock. It will be carried on a US satellite built by Surrey Satellite Technologies Limited (SSTL) as part of the USAF's space test program (STP-2) mission to orbit Earth. Once in orbit, DSAC will perform spacecraft navigation and global positioning system tasks. DSAC improves the spacecraft's navigation ability to distant destinations, gathering more data with higher accuracy. DSAC will be 50 times more accurate than the current best navigational atomic clocks.

4) The core of advanced robot technology utilizes the ability of robots to enhance human productivity. The risk of space missions can be reduced by integrating the capabilities of both humans and robots. For example, NASA is building a prototype of the detector to support the Resource Prospector project. The mission is aiming for the Moon, and could be launched in 2020. It will verify lunar exploration skills in order to determine the location and composition of large water ice that may be buried beneath the surface of the Moon. The project could lead to the construction of lunar 'gas stations' to extract water from the Moon's surface in order to make hydrogen and oxygen for rocket fuel.

5) On June 8, 2015, NASA conducted a Low-Density Supersonic Decelerator (LDSD) technology demonstration and verification mission at the US navy's Pacific Missile Range Facility (PMRF), which was of great significance for future Mars missions. The LDSD test used a saucer-shaped rocket powered inflatable deceleration vehicle to reach near space. It evaluated two key technologies for landing robots and support systems for Mars science and human exploration missions: the supersonic inflatable pneumatic decelerator (SIAD) and the Supersonic Ringsail (SSRS) parachute. The former is a large ring-shaped air brake that allows large payloads to land on Mars or other destinations with atmospheres.

4. How Opening is Promoting NASA's Innovations

NASA is renowned for performing complex and difficult space exploration missions, and its future projects are full of challenges and risks. Only by working to reinforce disruptive innovation can it find low-cost and low-risk ways into space. The following is a summary of NASA's efforts in recent years to improve its disruptive innovation capabilities. The core of these efforts is to open up to the public in order to gather ideas and accelerate the development of disruptive space technology.

4.1 Establishing organizations that specialize in disruptive technology

In order to improve its ability in disruptive technological innovation, NASA has also carried out institutional innovation, establishing a new department called the Space Technology Mission Directorate (STMD) in 2013. STMD will invest extensively in the development of bold, broadly applicable disruptive technology that the industry cannot yet tackle.

According to NASA regulations, STMD's mission is to improve America's space capability and NASA's ability to complete its missions through rapid R&D and the integration of technology. STMD's investment in space technology can enhance NASA's capability in terms of entering and traveling in space. It can also allow more efficient operation of satellites in Earth orbit and beyond, and help NASA to put systems into space more accurately, while improving survivability in deep space and on exoplanets. Thus far, STMD has carried out three technological R&D and innovation programs, namely the Small Business Innovation Research program (SBIR), Small Business Technology Transfer program (STTR) and Space Technology Research and Development (STR&D). These projects are open to the public.

The technological research and innovation program is funded by the Office of the Chief Technologist (OCT), and aims to integrate NASA's technology development and open innovation activities. The OCT is responsible for developing the NASA Space Technology Roadmaps and the NASA Strategic Space Technology Investment Plan, and also takes charge of technology transfer. In addition, the OCT maintains NASA's Technology Portfolio System (TechPort).

The Small Business Innovation Research (SBIR) Program and the Small Business Technology Transfer (STTR) Program support early-stage research and research into medium-maturity technology by US small businesses through competitive awards and contracts. SBIR and STTR provide research funds to high-tech small businesses to develop and commercialize new aerospace technologies that will help NASA to fulfil the needs of the US aerospace industry.

The Space Technology Research and Development (STR&D) Program is developing and verifying short and long term technological solutions that will enable NASA to perform missions more efficiently, at lower cost and with greater reliability. The project includes the following:

1) Early project investment incubation includes basic research, application research, and the development of early stage technology to establish a foundation for capacity innovation. The process is non-linear, and takes a great deal of time. Usually, early projects have no clear answers regarding whether they can achieve technological breakthroughs, significantly improve outcomes, or explore new paths. Therefore, a reasonable technology portfolio can only be considered after balancing investments in early stage projects, mid-Technology Readiness Level (TRL) projects, and technological demonstration and verification projects.

2) The Game Changing Development (GDC) Program advances space technologies that may lead to entirely new approaches for the Agency's future space missions and provide solutions to significant national needs. The program will focus efforts in the mid-Technology Readiness Level (TRL) range (3–5/6), generally taking technologies

from proof of concept through component or breadboard testing in an appropriate environment. The program employs a balanced approach that consists of guided technological development efforts and competitively selected projects from across academia, the industry, NASA, and government agencies. The program strives to develop the best ideas and capabilities irrespective of their source.

3) NASA's Technology Demonstration Missions (TDM) bridges the gap between scientific and engineering challenges and the technological innovations needed to overcome them, enabling robust new space missions. These include the Restore-L satellite service, Deep Space Optical Communications (DSOC), Deep Space Atomic Clock (DSAC), the Green Propellant Infusion Mission (GPIM), Solar Electric Propulsion (SEP), Laser Communications Relay Demonstration (LCRD), Landing Accuracy, Low-Density Supersonic Decelerator (LDSD), Composites for Exploration Upper Stage (CEUS), and In-space Robotic Manufacturing and Assembly (IRMA).

4) Within the Small Spacecraft Technology Program (SSTP), by investing in small spacecraft, NASA provides a fast and low-cost solution for in-orbit verification of new technologies. It focuses on the development of the Pathfinder Technology Demonstrator (PTD) project to verify the miniaturized electric propulsion system, electric thrusters and other innovative propulsion systems, and Hall Effect thrusters by iodine propellants.

5) The Centennial Challenges scheme offers incentives to support innovative solutions with the goal of backing NASA's future exploration and advanced operations near and beyond the Moon. Projects include sampling return robots, the Cube Quest Challenge for exploring space missions beyond Moon with small satellites, the 3D-Printed Habitat Challenge, and a Mars air vehicle.

6) The Flight Opportunities program provides access to space for new technologies to complete technical flight verification activities through existing commercial sub-orbital flight capabilities. The project first selected technologies with advantageous application prospects from aerospace companies, universities, and government agencies, and then provided experimental opportunities on commercial sub-orbital launch vehicles. This method provides more flight test opportunities for new technologies, and also offers funding for the development of related suborbital vehicles. The Flight Opportunity Program has now selected 160 technology payloads, including robots and 3D printing.

The STR&D plan solicits project proposals from individuals and institutions both inside and outside of NASA through the NASA Innovative Advanced Concepts (NIAC) program. IT is not limited to NASA.

4.2 Promoting technological innovation in commercial aerospace

NASA's policy to promote commercial space travel has created a number of innovative commercial space companies, and has also enhanced its innovative capabilities in space exploration.

All of NASA's funds are invested by the US government. The Agency's budget application is approved by the President after being reviewed by congress. In the late 1960s, NASA's annual budget was as high as 43 billion US dollars (converted to the 2014 value) for the Apollo moon landing program, but averaged only about 17. 5 billion after 1972. For example, NASA's budget for the fiscal year 2007 was $16.8 billion, of which the cost of manned space travel in low-Earth orbit (dominated by the space shuttle) accounted for more than half. In early February 2016, NASA submitted its budget request for the fiscal year 2017 to congress. It was 19. 03 billion US dollars – 255 million less than the 19.285 billion approved in fiscal year 2016.

Comparing NASA's budget in 2017 with that in 2007, it is clear to see that its commercial space policy has played a very positive role in promoting its manned Mars exploration goals for 2030 or later. This policy has catalyzed the growth of the commercial space industry in the United States. Although NASA has not invested much, it has made considerable efforts to revitalize. This has helped it to lower the high operating costs of space shuttles and the dilemma of retirement, and has also ensured that it has the financial resources to continue developing the Orion spacecraft, Space Launch System (SLS) rockets, and corresponding ground systems so that it can transport astronauts into low-Earth orbit or beyond in the future. It will also help NASA to remain a world leader in terms of scientific exploration, geosciences, aerospace technology, and aeronautical research. Undoubtedly, NASA's recent success in interstellar exploration is closely related to the effectiveness of its commercial space policy.

In fact, only when the market is fully open or the monopoly is broken can disruptive technological innovation emerge. For example, before 2006there were only large companies such as Boeing and Lockheed Martin in the field of aerospace commercial transportation in the United States. After the full opening of the commercial space travel field in 2006, a number of companies emerged such as SpaceX, Orbital ATK, Sierra Nevada, Bigelow Aerospace, Blue Origin, and Virgin Galactic. These companies have made technological breakthroughs in the field, providing rockets (such as the Falcon series) and spacecraft (such as the Dragon) with the same performance as large companies, but with half of the transportation costs. Particularly impressive are the rocket reusability technology designed by SpaceX and Blue Origin, the 'lifting body' space shuttle technology from Sierra Nevada, and the Expandable Habitat technology from Bigelow Aerospace. Whether a new technology can produce a disruptive effect requires a long period of practice. It is too early to draw a final conclusion on the innovations of these commercial space companies, but they are clearly paving a new path for space exploration.

NASA's commercial space policy has also led to new applications for the products it has developed in the past. Technologies that have accumulated in NASA's labs for a very long time can also be transferred into real engineering systems. Without the Agency's commercial space policy, these achievements would still be on the shelf. Moreover, the emergence of a large number of innovative and dynamic commercial space companies has also had a major impact on NASA's old system, and has greatly stimulated innovation among employees

4.3 Strong Support for Social Innovation

Innovation is now no longer the exclusive province of the government, scientific research institutions, and enterprise-led R&D centers. Increasing social forces can now participate in what has become an era of mass innovation. This leads to another innovation mode – social innovation. Unlike traditional institutional innovation, it involves the public. Open and transparent from top to bottom, it is a social division of work and decision-making. Social resources are used more effectively, and the diversification of participants can provide more creative possibilities from a wider variety of angles. The risk of failure can be avoided, the cost of innovation can be reduced, and the cycle period can be cut down.

NASA believes that the public is resourceful and capable of solving complex problems. It therefore uses approaches such as prizes and crowdsourcing to persuade public researchers to participate in open innovation activities.

The importance of social innovation can be illustrated by NASA's solution to solar flare prediction.

Accurate predictions of solar flares have plagued the Agency for years, as eruptions have a significant impact on space, Earth, and humanity, especially astronauts on space stations. However, after more than 30 years of data accumulation, this problem remained unsolved. In 2009, NASA released this challenge to a network platform called InnoCentive – an intermediary agency that specializes in solving scientific problems with a so-called crowdsourced innovation model. Unlike traditional innovative institutions, InnoCentive is an advocate of the 'non-certified doctrine'. Participants do not needed to have a doctorate or work in a laboratory, and there are no disciplinary or field constraints. Ultimately, the person who solved NASA's dilemma was not a renowned space physicist, but a retired radio frequency engineer living in a small town in New Hampshire. His name was Bruce Cragin, and he had studied the theory of magnetic reconnection in depth. His solution dramatically improved the accuracy of predictions, giving a prediction capability of up to eight hours with 85% accuracy.

This attempt by NASA to cast its net more widely gave it a taste for social innovation. It now cooperates with three open innovation platforms including InnoCentive, through which it has solved a number of challenging problems such as the preservation of food in space, solar activity prediction, and the non-invasive measurement of human intracranial pressure. Undoubtedly, social innovation provides a broad new way for the realization of disruptive technology and innovation for space.

In addition, in order to improve the aerospace literacy of social platform users, NASA has increased its investment in the popularization of space science. Its budget for this endeavor in fiscal year 2017 was as high as $100 million. No less, NASA's film *The Martian* – a collaboration with 20th Century Fox – premiered worldwide in October 2015, and played a significant role in promoting future Mars travel. Public support for space exploration is absolutely vital.

China's space industry has made unprecedented achievements, but there is still a major gap compared with America's innovation capabilities. The only way to remedy this situation is to strengthen innovation, particularly of the disruptive technological variety. In the 13th Five-Year Plan, mass entrepreneurship and innovation are new engines for promoting the sustainable development of China's economy. This policy has augured a series achievements in the Internet field, and is now entering the aerospace sector. The main lesson to learn from NASA in developing disruptive technology is to have an open mind. Through this method, NASA has broken the restrictions of its own culture, and has made full use of public resources to accelerate the innovation of disruptive space technology.

References

[1] Clayton M. Christensen, Joseph L. Bower. 'Disruptive Technologies: Catching the Wave' [J]. *Harvard Business Review*, 1995.

[2] Clayton M. Christensen. *The Innovator's Dilemma: When New Technologies Cause Great Firms to Fail* [M]. Boston, Massachusetts, USA: Harvard Business School Press, 1997.

[3] You Guangrong, Liu Changli, Wang Zhiyong. 'Paying Attention to the Development of Disruptive Technologies' [N]. *Science and Technology Daily*, 2013–07–22.

[4] Huang Zhicheng. 'Technological Innovation from the Perspective of STS' [J]. *International Technology and Economics Research*, 2006 (4): 24–28.

[5] Huang Zhicheng. 'Disruptive technology: a breakthrough in scientific and technological innovation' [N]. *China Youth Daily*, 2015–11–09 (2).

[6] Huang Zhicheng. 'Rational treatment of disruptive technologies' [N]. *China Youth Daily*, 2016-09-19 (2).

[7] Huang Zhicheng. 'Pre-research on new weapons in the United States' [N]. *China Youth Daily*, 2013-12-09 (2).

[8] Michael Belfiore. *The Department of Mad Scientists: How DARPA Is Remaking Our World, from the Internet to Artificial Limbs* [M]. Group translation. Beijing: Science Press, 2012.

[9] Defense Advanced Research Projects Agency. *Innovation at DARPA* [R/OL]. www.darpa.mil/attachments/DARPA_Innovation_2016.pdf. 2016.

[10] Defense Advanced Research Projects Agency. *Breakthrough Technologies for National Security* [R/OL]. www.darpa.mi/attachments/DARPA2015.pdf.2015.

[11] Chen Ao, Liu Xielin. 'Where does breakthrough technology come from?—a literature review' [J]. *Studies in Science*, 2011, 20 (9): 1281–1290.

[12] Jackie Fenn, Mark Raskino. *Mastering the Hype Cycle: How to Choose the Right Innovation at the Right Time* [M]. Gartner, Inc., 2008.

[13] United States Government Accountability Office. *Report to Congressional Committees. Key Factors Drive Transition of Technologies, but Better Training and Data Dissemination can Increase Success* [R/OL]. www.gao.gov/assets/680/673746.pdf.2015.

[14] Nie Haitao, Sang Jianhua. 'Demystifying the Skunk Factory (Part 2)' [J]. *Aviation World*, 2015 (8): 2–38.

[15] Du Hongliang. 'The Main Features of Silicon Valley as a Global Science and Technology Innovation Center' [J]. *Global Science, Technology, and Economy Outlook*, 2016, 31 (3): 46–50.

[16] Liu Dongxia. 'Analysis of American SME Innovation and Entrepreneurship Management Service Organizations' [J]. *Technological Innovation and Productivity*, 2016 (10): 10–16.

[17] Tyler Cowen. *The Great Stagnation* [M]. Tu Zipei, Trans. Shanghai: Shanghai People's Publishing House, 2015.

[18] Sun Jingfen, Yuan Jianhua, Zhao Yan, et al. 'Analysis of the implication, classification, and display methods of disruptive space technology' [J]. *International Space*, 2016 (7): 34–40.

[19] Huang Zhicheng. 'How NASA promotes disruptive technological innovation with openness' [J]. *Space Exploration*, 2016 (5): 30–34.

[20] Huang Zhicheng. 'How mass innovation has opened a new era for aerospace' [J]. *Military Industry Culture*, 2016 (5): 41–43.

[21] Liu Yingguo. 'Research and Thinking on the Characteristics of NASA's Open Innovation' [J]. *Satellite Applications*, 2016 (9): 18–26.

[22] E. J. V. D. Veen, D. A. Giannoulas, M. Guglielmi, T. Uunk, D. Schubert. 'Disruptive Space Technologies' [J]. *International Journal of Space Technology Management and Innovation*, 2012, 2 (2): 24–39.

CHAPTER 3

BREAKTHROUGHS IN CIVIL-MILITARY INTEGRATION

Space technology is applied in both the civil and military fields, meaning that it is an important area for the in-depth development of civil-military integration. This form of integration has become an inevitable development direction global space travel, and an important symbol of the new era. Commercial space travel has thus become a breakthrough in the development of civil-military integration.

1. A New Phase of Civil-Military Integration

A revolution in the sci-tech, industrial, and military fields is accelerating around the world. In order to seek early advantages in the intense international competition, major countries are expanding the depth and breadth of civil-military integration, trying to shape a new growth point in their military capability and comprehensive national strength. The development of global civil-military integration is entering a new stage that is characterized by innovation.

To handle the new wave of global civil-military integration, China should strengthen its sense of urgency, adapt to new trends, grasp new features, innovate its thinking, and strengthen its strategic response. It should also promote open innovation, back the development of civil-military integration, and strive to grasp the strategic dominance in military competition and comprehensive national strength.

On March 12, 2015, Xi Jinping – General Secretary of the Central Committee of the Communist Party of China, President of the State, and Chair of the Central Military Commission – attended the plenary meeting of the PLA delegation at the Third Session of the Twelfth National People's Congress, where he stressed that the development strategy for civil-military integration should be carried out in depth to strengthen and revitalize the army. Xi pointed out that 'to upgrade the development of civil-military integration into a national strategy is a major achievement of our long-term exploration of the coordinated development of economic construction and national defense, and a major decision made from the overall situation of national security and development.'

What are the reasons for this a new civil-military integration? First of all, the strategic game among leading powers is heating up, and the international strategic pattern is undergoing unprecedented changes. Since the global financial crisis, the economies of developed countries have slowed, and their overall strength has declined. However, some developing nations, especially emerging economies, are rising up. The struggle for global dominance and regional order is fierce. Competition and cooperation among major powers are intertwined. Due to the proliferation of nuclear weapons, ethnic contradictions, religious conflicts, and geopolitical uncertainty, the risk of war is increasing.

Second, after the global financial crisis, the world economy needed to recover. The history of world economic development shows that the emergence of new technologies, industries, and products is vital for recovery after a crisis. In recent years, the rapid development of civil-military integration industries around the world has become a new momentum to stimulate economic recovery.

Finally, the technological and industrial revolution is accelerating development, with extremely active global scientific and technological innovation. High-tech industries such as artificial intelligence, robots, unmanned systems, intelligent manufacturing, 3D printing, the Internet of Things, Big Data, new energy, new materials, and biotechnology have developed rapidly, creating a new era of civil-military integration across the world.

The new technological revolution has the following characteristics. First, the origin and source of high-tech innovation is shifting to the civil sector. Historically, military demand was driving force behind the technological revolution. New technologies often emerged from military demand, and were first applied in the military field. This tendency is now undergoing profound changes. The driving force behind high-tech development has shifted from military to civilian use, and the civilian sector has replaced the military sector as the trail blazer in the development of new technologies. Second, many new technologies have made breakthroughs in the civilian field, showing great prospects for military applications. High technology that can be applied in the military field is now developing rapidly. The 'human-machine war' in March 2016 attracted worldwide attention, showing the infinite potential of the application of artificial intelligence in the military field. Third, the boundary between military and civilian technology is increasingly blurred, and both technologies are becoming integrated. As high technology is becoming dual-use technology, it is difficult to distinguish which technologies are exclusively for military or civilian use.

In addition, with the evolution of modern weapons and equipment, the development and cost of technology is rising. In order to maintain prosperity, a major power must maintain a balance between safeguarding national security and improving public welfare, insisting on the coordinated development of national defense and economic construction. It is necessary to adopt the strategy of deep civil-military integration.

Clearly, civil-military integration is an inevitable trend that accords with the current international situation and the law of scientific and technological development. It has become an important guarantee for major countries to maintain sustainable development.

On June 20, 2017, Xi Jinping (General Secretary of the Central Committee of the Communist Party of China, President of the State, Chair of the Central Military Commission, and Director of the Central Committee for Civil-Military Integration and Development) chaired the first plenary meeting of the Central Committee for Civil-Military Integration and Development. In his speech, he stressed that China should strengthen centralized and unified leadership; implement an overall national security concept and military strategic principles within the new situation; highlight problem orientation; strengthen top-level design; strengthen demand integration; coordinate incremental stock; and synchronously promote systematic and mechanism reform, system and element integration, and the construction of systems and standards. He stated the necessity of accelerating the formation of deep civil-military integration across all elements and multiple fields, with high benefits. In this way, China will build a national strategic system and capacity for civil-military integration.

Since the founding of the People's Republic of China, civil-military integration has achieved remarkable results, moving from small-scale to large-scale. Institutional mechanisms have been improved, and policies have been introduced. Civil-military integration of weapons and equipment, national defense mobilization, personnel training, and military support systems have become increasingly sound. Civil-military integration in emerging areas such as information, space, and oceans has been steadily promoted. Industrial bases for civil-military integration and high-tech parks have been established, while the in-depth development and construction of civil-military integration has entered a period of acceleration.

Civil-military integration is an upwardly spiraling process from quantitative change to qualitative change, then from qualitative change to new quantitative change. China's civil-military integration has now entered a new stage of development, moving from initial integration to deep integration. In order to promote the deep development of civil-military integration at this new starting point, China must fully understand its historical orientation, accurately handle its current characteristics, solve its outstanding contradictions and problems, and strive to form a comprehensive, multi-field, and highly beneficial pattern for the deep development of civil-military integration.

2. Progress and Restrictions for Military aerospace

In recent years, major nations have accelerated their development in the field of space warfare, and the militarization of space has accelerated significantly. This has aroused widespread concern about security issues around the world.

2.1 Changes to International Military Aerospace Strategies

Science and technology in the aerospace sector has advanced considerably, and military theories are being constantly innovated to stimulate the evolution of warfare. All military powers regard space power as a strategic means, with equal importance to nuclear power. In recent years, the United States military has been leading the global military revolution, and has spared no effort to develop space power. The US space force were the first to put military satellites into use, which played an important role in supporting battlefield information. The US army is highly information-based. All military operations rely on sufficient information support, and almost all weapons are more or less dependent on the support of space-based information systems. Looking back to previous overseas military operations of the US military, the operational effectiveness of most precision-guided weapons could not have been achieved without the support of space-based information systems. The US military is now building the Prompt Global Strike (PGS) program, which will rely heavily on space-based information systems. Meanwhile, battlefield response time will be substantially reduced.

The core military goals of the US space force are to obtain the power of space–domination on the battlefield, ensure the free and accessible use of space in wartime, and take full advantage of space-based information systems in providing both information and communication support. As a result, the US space force's mission will expand from entering and using space to controlling it. The use of space is accelerating its penetration from the strategic to the tactical level. Countermeasure technology to control space has attracted additional attention, with a growing trend towards militarization.

As a major national interest, space plays an extremely important strategic role in national security, economic and social development, technological progress, and the enhancement of political influence. With space becoming increasingly congested, confrontational, competitive, multi-polar, and globalized, major countries have issued many important military aerospace strategies and policies to plot the future development of military aerospace. In this way, they hope to accelerate the updating of equipment and enhance the military application capability of space systems. Meanwhile, China is supporting the development of commercial space, and promoting the expansion of commercial space to military applications. Developing commercial space has become a breakthrough in the in-depth development of civil-military integration.

Driven by the US Department of Defense, the Clinton administration introduced the first National Space Policy, in 1996 which stated that 'Peaceful purposes allow defense and intelligence-related activities in pursuit of national security.' This document opened the door for the US military to develop its space warfare capabilities. In January 2001, the Commission to Assess United States National Security Space Management and Organization (NSSMO), led by Congress and then US Defense Secretary Donald Rumsfeld, submitted a report to Congress: 'If the US is to avoid a "Pearl Harbor in space", it needs to take seriously the possibility of an attack on the US space system.' In 2006, the Bush administration issued a strongly upgraded version of the National Space Policy, which pointed out that 'if necessary, the United States will … deny its adversaries the use of space capabilities for US national interests.'

Launched by the Obama administration in June 2010, the new version of the National Space Policy proposed to strengthen military aerospace cooperation with allies, providing a legal basis for the implementation of joint space operations between the US military and allied forces. Six months later, the US Department of Defense (DoD) issued the first National Security Space Strategy (NSSS), which stipulated three major tasks for the US military in the field of space warfare: 'preventing and deterring aggression against space infrastructure that supports US national security'; 'preparing to defeat attacks', and 'operating in a degraded environment'. This document claimed that the United States was committed to establishing 'guidelines' for space infrastructure operations and satellite launches for international society in the future, following the 'norms of behavior' of countries regarding space. On January 5, 2012, then President Barack Obama released a military strategy report entitled 'Sustaining US Global Leadership: Priorities for 21st Century Defense', which stressed that the US was determined to maintain its 'military superiority' despite the reduction of defense spending while shifting its military focus to the Asia-Pacific region. The report pointed out that 'to enable economic growth and commerce, America will work in conjunction with allies and partners around the world to seek to protect freedom of access throughout the global commons to those areas beyond national jurisdiction that constitute the vital connective tissue of the international system'. The term 'commons' refers to 'areas that do not belong to any single country and can be used by most countries in the world'. Specifically, it includes the ocean, space, sky, and network. Controlling the 'commons' means to realize the basis of US military hegemony. However, the word 'controlling' here does not mean that other countries cannot use 'commons' in peacetime. Its real sense is that the United States can obtain greater military use from them than other countries.

In February 2015, then president Barack Obama unveiled a new National Security Strategy that was more explicit about America's response to the space threat than previous ones: 'We will also develop technologies and tactics to deter and defeat efforts to attack our space systems; enable indications, warnings, and attributions of such attacks; and enhance the resilience of critical US space capabilities'.

On June 28, 2011, the Obama administration released the National Space Policy of the United States of America, which stated that it would: 'enhance the protection and resilience of selected spacecraft and supporting infrastructure; develop and exercise capabilities and plans for operating in and through a degraded, disrupted, or denied space environment for the purposes of maintaining mission-essential functions; and develop transparency and confidence-building measures'. It also stated that 'the United States will pursue bilateral and multilateral transparency and confidence-building measures to encourage responsible actions in space, and the peaceful use thereof. The United States will consider proposals and concepts for arms control measures if they are equitable, verifiable, and enhance the national security of the United States and its allies'.

The US space security strategy is clearly evolving from an emphasis on offense to a combination of deterrence and defense, having realized the need to return to the negotiating table. To this end, the US military continues to increase its investment in Space Situational Awareness (SSA) and space defense. In August 2013, the US Air Force Space Command (AFSPC) issued a white paper entitled 'Resilience and Disaggregated Space Architecture', proposing that in order to identify threats from potential adversaries' space control approaches (including fractionation, functional disaggregation, hosted payloads, multi-orbits, and multi-domains), the existing satellite constellations and space architecture will become 'resilient' and 'disaggregated' in order to yield greater survivability. In addition, the US military has vowed not to give up the development of anti-satellite weapons, and will retaliate by all the means if its satellites are attacked.

In May 2016, the US Air Force Space Command (AFSPC) released a new version of 'The Commander's Strategic Intent'. This document stressed the priorities and overarching strategic concepts of AFSPC, including building resilience capacity; emphasizing multi-domain combat effects; developing agile information superiority; emphasizing intelligence, surveillance, and reconnaissance (ISR); and improving buying power. This will affect the procurement, development, and application of future US military satellites. In the FY2017 Defense Budget of the United States, the investment in space-based systems was 7.1 billion US dollars, which was equal to the FY2016 figure. The US will now focus on the procurement of an Evolved Expendable Launch Vehicle (EELV), the development of a Next-Generation Weather Satellite and an Advanced Extremely High Frequency (AEHF) satellite system, the maintenance of a Space-Based Infrared System (SBIRS) system, and the development of an Operational Control System (OCX) for the Global Positioning System (GPS).

Russia is developing in all aspects of space capability, and is striving to consolidate its leading position in the field. In July 2015, President Putin approved the Federal Law on Russian Federal Space Agency (ROSCOSMOS) to strengthen the management of the aerospace industry by integrating the government and enterprises. This overturned the ongoing reform measures of separating the government from enterprises in the Russian space field, and sought to intensify and concentrate management. In August 2015, the Russian Airspace Forces began to take on combat duty tasks, and were responsible for unified command and management of air operations, air defense, and anti-missile forces. The formation of the Russia Airspace Forces has integrated forces and resources in many fields, and has established an airspace combat system for both attack and defense.

2.2 The Restrictions of Military Aerospace

The development of military aerospace is currently accelerating, but it is constrained by strategy, technology, and funding.

First, military aerospace is strategically constrained by the balance of major powers. Aerospace has changed the traditional concept of offense and defense. Offenses launched in the information space are heavily concealed, the cost of attack is small, and the damage can be great. At the same time, it is almost impossible to achieve perfect defense. In 2007, a country launched a ground-based anti-satellite missile that blew up an abandoned weather satellite. In 2008, the United States shot down a reconnaissance satellite with a Standard Missile-3 (SM-3) interceptor. This test proved that the US's anti-missile system potentially had anti-satellite capabilities. These two launches had a profound impact on America's space security strategy as well as global space security

The development of anti-satellite weapons means that the space arms race is bound to intensify, and is closely related to the progress of arms control in space. In the late 1970s, the USA and the USSR conducted three rounds of negotiations on anti-satellite arms control. There was some progress, but the two sides failed to reach an agreement on anti-satellite weapons for various reasons.

There are several treaties and agreements that have limited the placement of certain weapons in outer space and have provided some protection for satellites. Major treaties include the Limited Test Ban Treaty (1963), which banned all nuclear tests in the atmosphere, in space, and underwater. The Outer Space Treaty (1967) provided the basic framework for international space law, including the stipulations that 'states shall not place nuclear weapons or other weapons of mass destruction in orbit', and 'the exploration and use of outer space shall be carried out for the benefit and in the interests of all countries, and shall be the province of all mankind.'

To deal with the threat of anti-satellite weapons, the US has two options: space weaponization and risk avoidance. The former refers to the persistent deployment of weapons systems in space to prevent attacks; the latter means to focus on reducing vulnerabilities and minimizing the negative impact if US satellites are attacked by other countries. Despite the pros and cons of both options, most American experts seem to advocate risk avoidance over space weaponization. They believe that the US can gain more access to space by reducing its military dependence on space assets and 'leading the international community to work out a code of conduct or more effective international laws governing space activities', as well as creating conditions to keep space 'clean' in order to benefit America's economy, politics, and national security. As a result, the United States has made many adjustments to its space security strategy.

The Chinese government believes that the issue of the space arms race is one of the most urgent and prominent in the field of international arms control and disarmament, and has the greatest impact on international peace and security in the 21st century. It should be one of the highest priorities on the agenda of the General Assembly on Disarmament, which is the sole multilateral disarmament negotiating body.

Secondly, technical obstacles exits. The aforementioned idea of the resilient structure of the space system proposed by the US Air Force Space Command has not progressed smoothly. It is blocked by vested interest groups, and also encounters a series of technical obstacles. In February 2008, the Defense Advanced Research Projects Agency (DARPA) awarded Phase I of the F6 system contract to four contractors for one year. F6 is an acronym of the six words 'Future, Fast, Flexible, Free-Flying, and Fractionated spacecraft' and the goal of the project was to identify a new way to accomplish space missions. The satellite system of the F6 program would have replaced the fully-featured large satellite with a small satellite group for special missions, and would simply have updated or changed its mission by adding new satellites to the constellation. However, the project was suspended in 2013 after technical problems with its implementation.

The final constraint is economic affordability, which is a key factor for the development and design of military aerospace systems. For most systems, the cost of use, maintenance, and improvement is higher than the cost of purchase. Therefore, the lifetime cost must be considered during the phases of technological development and demonstration. China should also pay attention to the overspending that has become a common phenomenon in the procurement of modern weapons and equipment. According to statistics from American surveys, the actual worth of US defense contracts increased by 200% on average in the 1950s, 400% in the 1960s, 100% in the 1970s, and 40% in the 1980s. In addition, the cost of space-based information systems remains high, and has continued to suffer over-expenditure. The total cost of the Space-Based InfraRed System (SBIRS-high) constellation – which consists of five satellites and replaced the Early Warning Satellite system's so-called Defense Support Program (DSP) –increased by 150% from 3.9 billion US dollars in 1996 to 10 billion in 2004. Finally, this program had to be suspended by the US Air Force according to the Nunn-McCurdy Act passed by Congress, as the Department of Defense must report to Congress if the cost exceeds 25%.

In 2007, The Center for Strategic and Budgetary Assessments in the United States released a report that estimated that the lifetime cost of running space-based kinetic energy interceptors for 20 years was between 29 billion and 290 billion US dollars. The lifetime cost of a space-based laser running for 20 years was 128 billion to 196 billion. However, the technical risk of developing space-based lasers is much higher than that of developing space-based kinetic energy interceptors. Otherwise, the ground-based, sea-based, and air-based missile defense systems being used and developed in the United States would cost much less, with a lifetime cost of between 15 billion and 80 billion US dollars. For most purposes, the report concludes that ground-based systems are more capable and cost-effective than space-based systems. For space-based ground-attack weapons, the report also argues that the cost is much higher than for ground-based or sea-based weapons systems. The reason for the high cost of space-based systems is that the current launch cost remains high, as well as higher maintenance cost. This is due to the vulnerability of in which space-based systems in terms of structure and electronic subsystems in harsh environments, especially for laser systems.

In order to overcome these constraints, in-depth civil-military integration is an inevitable strategic choice. It can accommodate military and civilian purposes, and can also overcome the technical obstacles of military aerospace systems by civil space technologies and commercial markets, which greatly reduces costs.

3. Breakthroughs in Civil-Military Integration

On June 22, 2017, President Xi Jinping, as Chairman of the Central Military Commission, made a speech on a visit to a military base in Shanxi Province. He stressed that efforts should be made to deepen civil-military integration; to seize the CPC Central Committee's strategic opportunity to promote civil-military integration; to accelerate the pace of exploration and practice; to find the right path for deep integration in technology, industry, facilities, and personnel; and to strive to ensure the leading position of civil-military integration in the space arms field.

3.1 Aerospace: at the forefront of civil-military integration

Aerospace technology is dual-use. It can be used for both civilian and military purposes, and also as an aerospace product. For example, launchers based on rocket technology can launch both civilian and military satellites. Spacecraft based on space shuttle technology can be used to send shuttles into orbit for both human and cargo delivery, while X-37B can be used only for military purposes. This is especially true for satellite technology. For instance, navigation satellites can be used to code differently for military and civilian purposes. Many commercial remote-sensing satellites benefit from military reconnaissance satellite technology. The distinction between military and civilian weather satellites is harder to draw.

Therefore, aerospace is an important area for the deep development of military and civilian integration. This integration has become an inevitable direction for the global space industry, and has also become an important symbol of aerospace in the new era.

The United States leads the world in civil-military integration. After the Cold War, American defense spending was reduced due to the economic slowdown, and the economic benefits of military projects remained concealed. In the late 1990s, the US government identified economic development as an important part of its new national security strategy, and began to discuss how to combine military requirements with civilian technologies to promote it. In recent years, with the consistent reduction of national defense funding, the United States emphasizes the use of technological innovation in the civilian economy to realize the leap-forward development of sci-tech for national defense.

America's space industry is divided into three major parts: public, commercial, and military aerospace. Civil-military integration in space includes the integration of military and public aerospace, as well as the integration of military and commercial aerospace. There is no difference in the fundamental industrial nature of these three parts; only the application objects and development goals are different. Therefore, an integrated development mode can give full play to their respective characteristics and realize complementary advantages.

3.2 The Integration of Military and Public Aerospace

As NASA is responsible for public aerospace activities in the US, the integration of military and public aerospace is only possible in cooperation with the military.

On July 29, 1958, then president Dwight D. Eisenhower signed an act establishing the National Aeronautics and Space Administration (NASA), replacing its predecessor, the National Advisory Committee for Aeronautics (NACA), which was responsible for developing and implementing the civil space program and carrying out aerospace research. NASA inherited NACA's tradition of developing dual-use aerospace technologies.

In its early civil space activities, NASA paid great attention to the development of dual-use space technology, especially when developing space transportation systems. Many ballistic missiles that were developed by the military are also used in civilian space launches. For example, Project Mercury used Atlas missiles while Project Gemini used Titan missiles, both of which came from the USAF. This cooperation was interrupted during the Apollo Moon Landing Program, but afterwards, NASA and the military placed their hopes in the space shuttle for low-orbit space delivery. The military even abandoned the development of carrier rockets as a result. However, the high cost of space shuttle transportation forced the USAF to begin developing new launch vehicles (EELV) and form the United Launch Alliance (ULA), which operates the Delta II, Delta IV, and Atlas V launch vehicle systems.

NASA and the military have long worked closely on advanced space delivery systems and hypersonic technologies. In the mid-1980s, NASA implemented the National Aerospaceplane (NASP) program with strong support (80% funding) from the US Air Force. Since then, NASA has conducted a smaller flight test for the X-43A. After this test, according to the National Aerospace Initiative (NAI) that was jointly developed by NASA and the military, the main research into hypersonic technology was transferred to the United States military. The flight test of the X-51A was conducted with a hypersonic missile as the background and flight test of other hypersonic vehicles. Another example of the NASA-US military relay was the development of the renowned unmanned space shuttle, the X-37B. It was originally developed by NASA as the X-37, but was later handed over to the military, which changed its name.

After the decommissioning of the space shuttle, NASA hoped to develop commercial launch vehicles to transport cargo to the International Space Station. However, due to the successive crashes of the Antares-130 rocket from the Orbital Sciences Corporation and the Falcon 9 rocket from Space X, the next cargo mission could only use the ATLAS V rocket to launch the Cygnus cargo spacecraft from the Orbital Sciences Corporation. Boeing's CST-100 manned spacecraft was also launched by the ATLAS V rocket, and the first unmanned flight test of NASA's own Orion spacecraft was launched by the Delta IV rocket. Clearly, there is considerable scope for cooperation between the military and NASA on launch vehicles. The USAF will also collaborate with NASA on a study of the next-generation upper stage propulsion system, which will replace the military's RL-10 engine.

In other areas, the military and NASA have worked together in depth, including for space surveillance, on-orbit services, networking, communications, and remote-sensing. It should be noted that the military and NASA also offer mutual convenience when using ground infrastructure. For example, wind tunnels at NASA's research centers were delivered to the military during the development of the X-37B and X-51A. Meanwhile, the USAF's hypersonic Wind Tunnel 9 at the Arnold Engineering Center was used by NASA to develop the Orion spacecraft.

3.3 Commercial Spaceflight: A Breakthrough for Civil-Military Integration

The rapid development of commercial space travel has made it into a breakthrough for the in-depth development of civil-military integration for the aerospace sector.

In-depth civil-military integration in aerospace started early in the USA, and has relatively complete laws and regulations. The development of military and commercial aerospace in the US is now not only balanced, but also complementary. The integration of military and business has become the norm within the aerospace industry in a variety of modes such as priority buyout, part time leasing, and mutual overlap.

In the context of the recent drastic cuts to defense budgets and changes in the space security environment and space architecture, the United States Government has put more emphasis on the deep integration of military and commercial activities, explicitly proposing to 'maximize the procurement and use of commercial aerospace to meet government needs (including military needs)', and encouraging new measures for civil-military integration in 'deploying more satellites or payloads on non-military platforms'. According to the Commander of US Strategic Command in the House Armed Services Committee testimony in February 26, 2015, 'the USA continues to partner with responsible nations, international organizations, and commercial firms to promote the responsible, peaceful, and safe use of space.'

As commercial aerospace has become a new force in global space activities, more countries' military and government are 'buying services' through commercial activities to promote the in-depth integration of military and commercial aerospace. The US military has now become the largest user of commercial remote-sensing data in the world. Private firms are encouraged to participate in the aerospace industry through strong policy support for commercial aerospace, to promote the construction of a strong base for the aerospace industry, and to meet the needs of national space security more effectively.

In 2003, the US Commercial Remote-sensing Policy was issued, which proposed a maximization of the practicality of US commercial remote-sensing space capabilities for filling imagery and geospatial needs for military, intelligence, foreign policy, homeland security, and civil users. The 2010 edition of the US National Space Policy stressed the 'purchase and use of commercial aerospace capabilities and services to the maximum practical extent when such capabilities and services are available in the marketplace and meet United States Government requirements.' It also suggested a 'modification of commercial aerospace capabilities and services to meet government requirements when existing commercial capabilities and services do not fully meet these requirements, and when the potential modification represents a more cost-effective and timely acquisition approach for the government'. From the highest level, the United States has clearly and carefully defined the division of military and commercial labor and mutual coordination.

The United States repeatedly stressed in the aforementioned National Space Policy that the efficiency of the national security system can be greatly enhanced by improving the interoperability and compatibility of national security systems across the combat and mission areas, and by ensuring that this interoperability and compatibility are integrated into aerospace systems. While making full use of civil and commercial aerospace resources, the military should also pay more attention to system security measures as well as application compatibility and flexibility. The US military is developing multi-frequency, flexible, and intelligent communication terminals to make better use of commercial communication satellite resources. For example, the Navy Multiband Terminal (NMT) can replace various existing naval satellite communication terminals for a more effective integration of the communication resources of military and civilian satellites. The US Navy's Commercial Bandwidth Satellite Program (CBSP) aims to improve the Navy's ability to communicate using commercial satellites by developing commercial satellite communication terminals for a variety of ships and carriers. The US military has developed the Distributed Common Ground System (DCGS), which aims to improve the processing, distribution, and utilization of information to achieve intelligent information sharing. The system can integrate the isolated smokestack-type Intelligence Surveillance and Reconnaissance (ISR) system into a common modular intelligence architecture that is capable of collecting, distributing, and processing a variety of ISR information, automatic analysis, and data fusion. In addition to jointly processing, distributing, and sharing data from military aerospace-based information systems for various military services, future commercial aerospace-based information systems will also be included.

While emphasizing openness and compatibility, the security and availability of the system is always the primary issue for the US military in promoting the development of civil-military integration. In order to ensure the safe military use of GPS and the sustainability of civilian use, the US military has adopted a variety of technical measures to improve the anti-jamming capability of the system. For example, GPS devices and mobile phones adopt adaptive zero-adjusting antenna technology to ensure the normal operation of navigation equipment of the US military. They add new military M codes, separate military and civilian signals, restrict the use of GPS by adversaries, and reduce the interference to civil signal reception. The US military has also extended anti-jamming protection measures from military satellites to commercial satellites, including commercial switching centers, network security, and terminal security. In addition, it strengthens the protection of important leased commercial satellite communication assets by purchasing Global Satellite Jammer Geolocation services. Meanwhile, commercial satellite companies will enhance the security and reliability of commercial terminals by adopting more secure links, and developing more powerful encryption software, acceleration software, and virtual private network software. For example, Iridium handheld devices adopt encryption algorithms approved by the National Security Agency of the United States, which can only be used and shared by the military and specialist users.

In order to make full use of commercial resources to meet the needs of military capabilities, the military has taken economic means such as directly subsidizing private companies, investing in the construction of systems, leasing bandwidth, and purchasing commercial services. One example is the US military's support of the Iridium System in its bankruptcy, which also played an important role in supporting US military communications in subsequent local wars. In recent years, the National Geospatial Intelligence Agency (NGA) has invested in commercial satellite firms through the NextView project to develop and produce high-resolution imaging satellites, which gives it priority to use the satellites. In addition, in order to make up for the shortage of military communication resources during wartime, the military has developed a series of plans to purchase commercial satellite communication services. The US Defense Information Systems Agency (DISA) and the General Service Administration (GSA) have launched an important procurement program called the Defense Information System Network Satellite Transmission Service-Global pact to enhance fixed satellite bandwidth forwarding capabilities, improve terminal and bandwidth combination subscription services, and create an end-to-end End solution. In terms of GPS management and application, the United States stipulates that while protecting national security and foreign policy interests, it should enhance its economic competitiveness and productivity, and provide standard positioning services free of charge.

4. New Trends in Civil-Military Integration in the American Aerospace Industry

Since 2010, the United States has launched a number of new initiatives in aerospace information systems, space security, and launch vehicles to integrate the military with the civilian community, including greatly improving the security of commercial satellite communications by replacing terminal modems, increasing attack warnings and protective measures, and exploring the ability to invest in all or part of a commercial communication satellite in exchange for access to the entire commercial satellite network of commercial firms. It has also attempted to put some operational control functions like the Wideband Global SATCOM (WGS) constellation under the routine responsibility of the business community, as well as reducing accidental collisions and eliminating intentional interference of satellite frequency by commercial satellite data. To improve situational awareness in space, an agreement for the sharing of satellite positioning data was reached in August 2014 between the US Strategic Command and Space Data Association (SDA) founded by commercial firms. The US Air Force, intelligence services, and industry are also working on sharing detection data and developing the so-called game changing ability to achieve full utilization between various military and civilian services using the Space-Based Infrared System (SBIRS). In May 2015, SpaceX was certified by the US Air Force to launch national security missions, including satellites of the National Reconnaissance Office (NRO). On May 1, 2017, SpaceX successfully launched a highly confidential NROL-76 reconnaissance satellite for the NRO, subordinated to the US Department of Defense by the Falcon 9 launch vehicle.

4.1 HoPS: A New Exploration of Civil-Military Integration in Space

In recent years, the US Air Force has quietly promoted the Host Payload Solution (HoPS), drawing many lessons for future development.

On July 10, 2014, the HoPS project signed an Indefinite Delivery, Indefinite Quantity (IDIQ) contract with 14 firms including major satellite manufacturers and operators in Europe and the United States, namely Boeing, Lockheed Martin, Airbus, Arianespace, Northrop Grumman, Raytheon, Harris, Inmarsat, Intelsat, Iridium, Société Européenne des Satellites (SES), and Space Systems/Loral (SSL). The 14 companies formed the Host Payload Alliance (HPA) in 2011. The IDIQ contract also included projects for Boeing and two other companies to carry a Tropospheric Emission: Monitoring of Pollution (TEMPO) payload for air pollution measurements in most parts of North America. Previously, when the US Air Force and the Alliance issued contracts in accordance with traditional business rules, it was difficult to reach an agreement on price and technical standards. After nearly two years of repeated bargaining, the two sides finally reached a consensus.

In 1976, Hughes' commercial satellites carried military communications payloads from the US Navy. NASA's payloads were also carried on many occasions. However, the US Air Force adopted this method for the first time, carrying its Commercial Hosted Infrared Payload (CHIRP) on the SES-2 communication satellite launched by the Ariane 5 rocket in September 2011. CHIRP is an early warning missile experimental sensor that tests the capability of the US Air Force's Wide-field Infrared Staring sensor to perform large-scale surveillance in space. After two years of in-orbit testing, a large amount of target and sensor performance data was obtained, providing an important basis for future design. Reportedly, the cost of the payload was only 25 million US dollars.

The US Air Force had planned to replace the Defense Support Program (DSP) satellite it had been building since 1970 with the Space-Based Infrared System (SBIRS), but progress was not smooth. SBIRS consists of a large dedicated satellite in Geostationary Orbit (GEO) and associated payloads on large satellites in High Elliptical Orbit (HEO). In May 2011, it successfully launched SBIRS-GEO's first dedicated satellite weighing three tons, nine years behind schedule. According to a report by the Government Accountability Office (GAO), the cost of each satellite was as high as 3 billion US dollars. Therefore, in order to complete the sensor performance test for SBIRS with less money, the US Air Force implemented the CHIRP program. The successful completion of this mission has boosted the US Air Force's confidence to continue carrying infrared space sensors, regular space sensors, and space environment sensors on commercial satellites. However, there are still doubts about carrying military communication payloads on commercial satellites. It is believed that these payloads bring high demands for satellite quality and power, and spectrum compatibility makes it difficult for them to adapt to military and commercial communication payloads on a single satellite platform. In this regard, the Australian Defense Force (ADF) took advantage of the boarding payload opportunity to carry the UHF payloads they had purchased from Boeing on the Intelsat 22 satellite. The satellite was built by Boeing under a contract for Intelsat. According to the Australian authorities, the UHF payload program uses a boarding payload option that saves some 40% of the cost of conventional dedicated satellite purchases and launches.

HoPS's governing body is the Air Force Space and Missile Systems Center (SMC) Development Program Command, which runs the US military aerospace system.

The HoPS program was designed to provide the US Air Force and other US government organizations with the capability to host government payloads on commercial satellites in order to accomplish their missions. The HoPS contract required the other party to build a fully functional in-orbit payload system along with equipment and interfaces for the integrated ground system that would send payload data to government end users. Multiple payloads are common in the satellite industry. For example, electronic components and sensor packages designed for various special missions are mounted on the same spacecraft, sharing launch, propulsion, power, and costs. The boarding of military payloads on commercial satellites not only saves money, but also greatly shortens the development time. Moreover, due to the decentralized deployment of military payloads, the damage resistance capability of the US Air Force's space architecture has been improved. In the next five years, 80 commercial communications satellites will be launched with a whole host of boarding payload opportunities, and the US Air Force is ready to invest nearly $500 million in the project. Therefore, boarding military payloads on commercial spacecraft is an innovative way to integrate military requirements into commercial satellite missions.

4.2 Space Surveillance by Military-Commercial Cooperation

According to Aviation Weekly's report on June 7, 2015, the US government reached a consensus that space assets (including commercial, civil, and defense satellites) were no longer safe in orbit. The Air Force and six top commercial operators are conducting the first test in a new project to improve data exchange between them.

By the summer of 2015, the Joint Space Operations Center (JSpOC) at Vandenberg Air Force Base in California had begun a six-month trial run of the Commercial Integration Cell (CIC). JSpOC is the integrated situational awareness and command & control center that ultimately reports to the US strategic command. Its purpose is to improve the computer interface between commercial operators and the military. The CIC includes personnel and resources from six major DOD commercial suppliers, namely Digital Earth, Eutelsat, Inmarsat, SES, Intelsat, and Iridium. However, other commercial operators will also benefit from the CIC and its role as the industry representative and contact intermediary within JSpOC. Commercial operators have long called for in-depth co-operation with the DOD and intelligence communities. The CIC is one of several initiatives by which the Pentagon hopes to improve its Space Situational Awareness (SSA), receive information technology support from JSpOC, and store data on the computers in the operation center.

JSpOC announced that it has now tracked 23,000 targets in orbit, and the number is expected to increase as two new Geosynchronous Space Situational Awareness Program (GSSAP) satellites are put into use to deliver intelligence. The GSSAP satellite will help the US military acquire Space Situational Awareness and respond to electromagnetic interference, and help other commercial satellite companies to understand the dangers of operating satellites. During the pilot phase, the CIC will run eight to 12 hours per day, and participants will be on call in case of emergency. Eventually, the CIC unit will operate 24 hours every day. During the pilot period, participants will establish a 'machine-to-machine' interface to share data between military and commercial users.

Operators hope to emphasize the concept and process of aerospace protection in a variety of scenarios from the most dramatic satellite collisions or attacks to conventional electromagnetic interference. The routine procedures of the US military for dealing with most space emergencies, such as satellite collisions, are in place. However, with CIC, procedures for dealing with satellite interference events will be more rigorous and may help determine if there are any risk factors by providing operators with more data related to abnormal events. The pilot program is expected to form a standardized process through which commercial satellite operators can carry out training.

Since the White House approved an additional $5 billion investment in aerospace protection in the fiscal years 2016–2020, the CIC has been one of the key initiatives highlighted by the Air Force. In fact, other projects will benefit from investments to improve the JSpOC mission system, which allows commanders to respond to unusual events in space, rather than just observe them, according to General John E. Hyten, Commander of the US Air Force Space Command. The US Air Force may also start a follow-up project to the Space-Based Surveillance System (SBSS) satellite to collect more data, increase existing space target tracking catalogues, and possibly purchase more GSSAP satellites.

References

[1] You Guangrong, Zhao Linbang. *The Development of Military-Civilian Science and Technology Integration: Theory and Practice* [M]. Beijing: National Defense Industry Press, 2017.

[2] Sun Xinjing, Li Dong, Han Zheng. 'Promoting the in-depth development of military-civilian integration in science and technology for national defense' [J]. *Science and Technology for National Defense*, 2016, 37 (3): 10–13.

[3] Liu Yong, Jiang Feitao, Hu Wenlong. 'How military-civilian integration drives industrial transformation and upgrading' [J]. *China Economic and Trade Herald*, 2016 (36): 39–41.

[4] Huang Zhicheng. 'Understanding and Thinking about Space Warfare' [J]. *International Space*, 2003 (6): 10–15.

[5] Huang Zhicheng. 'Anti-satellite weapons and space arms control' [J]. *Satellite Applications*, 2000 (3): 1–8.

[6] Huang Zhicheng. 'A Review of the Development of US Space Confrontations' [J]. *Satellite Applications*, 2006 (3): 7–14.

[7] He Yusong. *Research into Space Security* [M]. Shanghai: Fudan University Press, 2014.

[8]   Huang Zhicheng. 'New Trends in the US Space Security Strategy' [J]. *International Space*, 2015 (12): 12–22.

[9]   Huang Zhicheng. 'US military aerospace: at the crossroads' [J]. *Space Exploration*, 2015 (5): 26–33.

[10] Huang Zhicheng. 'How the United States Avoids a 'Pearl Harbor Incident' in Space' [J]. *Space Exploration*, 2015 (10): 22–29.

[11] Wu Qin. 'A review of the development of global military aerospace equipment and technology in 2016' [J]. *Military Digest*, 2017 (1): 28–31.

[12] Zhang Baoqing. 'Trends in the development of global military aerospace' [J]. *Space Exploration*, 2017 (1): 40–43.

[13] Huang Zhicheng. 'New trends in the integration of civil and military in American aerospace' [J]. Space Exploration, 2015 (8): 9–13.

[14] Huang Zhicheng. 'The in-depth development of US military-civilian integration' [J]. *Satellite Applications*, 2015 (11): 30–34.

[15] Xie Ping, Li Chengfang. 'Some Thoughts on the Development of Military-civilian Integration' [J]. *China Aerospace*, 2017 (3): 10–15.

CHAPTER 4

GLOBAL AEROSPACE ENTREPRENEURSHIP

The rising tide of global space entrepreneurship is now spreading from the United States and Europe to China, and examples can be used for reference by Chinese space entrepreneurs.

1. Elon Musk's Dream

As a symbol of the new space age, a new wave of space entrepreneurship is sweeping the world. The most iconic figure of this wave is Elon Musk, the founder of Space Exploration Technologies (or SpaceX) in the United States.

On December 8, 2010, SpaceX launched the Dragon spacecraft into low-Earth orbit using the Falcon 9 rocket from Launch Pad 40 at Cape Canaveral Air Force Base in the United States. The Dragon spacecraft returned to Earth shortly after 2 p.m., landing 800 kilometers in the Pacific Ocean off the coast of southern California. After two more test flights, the Dragon is scheduled to berth unmanned with the International Space Station. If the berthing is successful, it will deliver 12 cargo flights there. If this goes well, it will be replaced by a manned spacecraft, with an estimated capacity of seven astronauts per flight.

The Dragon's launch and return mark the first successful recovery of a spacecraft from low-Earth orbit by a start-up commercial aerospace company. This is a feat previously achieved only by the United States, Russia, China, and other major powers. Since then, the SpaceX and its so-called 'maniac' founder have entered public consciousness in China.

1.1 Musk's Cross-Sector Entrepreneurship

Elon Musk was born in South Africa in 1971. His father was an engineer and his mother was from Canada. His father inspired Elon's love of technology, and bought him his first computer at the age of ten. The young Musk taught himself how to program, and at the age of 12, he sold his first piece of commercial software – a space game called Blaster. In 1989, he went to Canada to study at Queen's University in Kingston, Ontario, and eventually transferred to the University of Pennsylvania in the United States, where he received a double bachelor's degree in finance and physics. When he graduated, Musk decided to leverage his talent in the three most promising areas: the Internet, clean energy, and space.

Musk earned his first major paycheck from the Internet. In 1995, he and his brother Kimbal founded Zip2, a software company that provides online content distribution for news organizations. In 1999, Compaq bought Zip2 for $307 million and 34 million shares. In March 1999, Musk and other shareholders co-founded X.com – an online finance and e-mail payment service company. Later, X.com entered fierce competition with rival firm Confinity, and the two companies agreed to merge in March 2000. Musk took over as CEO and became the largest shareholder of the new company. In June 2001, X.com was renamed PayPal and sold to eBay with 1.5 billion shares in October 2002. Musk had been the largest shareholder before PayPal was acquired, and now owns 11.7% of the shares.

While Musk and his wife Justine were on their way to Sydney, Australia for their honeymoon, his rivals organized a board meeting while he was still on the plane. Before the plane landed, all of Musk's authority had been removed, and Peter Thiel had been appointed CEO. This emergency forced the couple to cancel their honeymoon. When they went to Brazil and South Africa for a belated holiday in late December 2000 and early January 2001, Musk contracted a rare form of malaria, but miraculously survived.

In June 2001, perhaps due to this near-death experience, Musk grew tired of life in Silicon Valley. He moved to Los Angeles and began to re-discover his interest in space exploration. Los Angeles is a hotspot for development in the aerospace industry, and offers easy access to the US Air Force and NASA. A year later in June 2002Musk founded Space Exploration Technologies Inc., also known as SpaceX, whose near-term goal is to develop lower-cost launch vehicles and spacecraft to substantially reduce the cost of getting into space. In the long term, Musk wants to create a brand new space frontier.

Musk's next step has been to invest in clean energy, and he is founder and CEO of Tesla – an American company that produces and sells electric vehicles, solar panels, and energy storage equipment. Musk has stated that Tesla will strive to provide ordinary consumers with pure electric vehicles within their consumption capacity. Tesla's vision is to accelerate the global shift to sustainable energy. In 2008, it released its first automotive product, the Roadster – a two-door sports car. In 2012, it released its second automotive model, Model S – a four-door electric luxury coupé. In September 2015, it delivered its third automotive product, Model X – a luxury pure electric SUV. In July 2017, Tesla launched its latest automotive model, the Model 3.

In 2008, Elon Musk became a shareholder in the new energy company Solar City. The company aims to change the way energy is produced in order to solve energy problems for average American families and make the planet cleaner. Solar City has installed solar panels into thousands of households, creating large distributed facilities, and is now the largest solar service provider in the United States. On November 17, 2016, Tesla acquired Solar City, becoming the only vertically integrated energy company in the world. It now provides customers with end-to-end clean energy products including electric vehicles, energy walls, and solar roofs.

Musk proposed a Hyperloop super train program in May 2013, and unveiled the initial design for the transportation project on August 12 that year. Hyperloop is a kind of transportation vehicle built around the core concept of 'Vacuum Pipeline Transportation'. In May 2016, Musk founded a new company called Hyperloop Transportation Technology (HTT) based on a passive magnetic levitation system developed by the Lawrence Livermore National Laboratory. The company has signed an exclusive agreement with this national laboratory. In the magnetic environment of a tunnel, the capsule can float before being 'thrusted' by the power supply and accelerate to just below the speed of sound. This also means that the operating cost of passive magnetic levitation systems will be greatly reduced. By 2020, HTT plans to test an 8 km route outside Los Angeles. If successful, longer Hyperloops could be built around the world.

1.2 Musk's Martian Dream

Elon Musk began to look to Mars in the late 1980s. At that time, he was a pupil at a boy's high school in Pretoria, South Africa, where he enjoyed launching model rockets and began fantasizing about colonizing other planets. His interest in space waned after he moved to Canada, and only resurfaced when he arrived in Los Angeles. At first, Musk had no idea what he would do in space, but after buying a second-hand Soviet space flight manual, he began to consider how space travel could change the world.

At this time, the Mars Society – a non-profit organization dedicated to exploring and colonizing Mars –issued a $500 request for donations to active members. Robert Zubrin, the President of the society, was surprised to discover that a stranger named Elon Musk had sent him a $5,000 check. Zubrin looked through Musk's profile and found that he was a very wealthy young man. He put him on the VIP table with renowned film director James Cameron and a NASA scientist called Carol Stoker at his fundraising dinner. Zubrin persuaded Musk to join a program called Translife, which would send mice into low-Earth orbit and spin their containers fast enough for them to simulate life and reproduction in the gravity of Mars. Musk also joined the Mars Society.

Although initially enthusiastic about the Translife program, Musk soon had a more ambitious idea than sending mice to Mars. His ultimate wish is not to colonize Mars, but to revive public enthusiasm for science, space exploration, and technology. After browsing the NASA website over and over again, he found that the current human space program was far from satisfactory. This strengthened his determination to revolutionize the seemingly sluggish space exploration program.

Musk's ambition, as well as his exploration of the aerospace industry, led him to resign from the Mars Society and establish his own Mars Life Foundation later that year. Soon after, the Foundation convened a working meeting with James Cameron, engineers and scientists from JPL, and Michael Griffin, CTO of Orbital Science, who became the director of NASA four years later.

Experts soon worked out a new plan to grow plants on Mars, called the Mars Oasis, for which Musk provided around 30 million US dollars. This meant that Musk had to buy a rocket that was cheap enough to be used to put an automatic greenhouse on Mars. He finally decided to go to Russia to buy a Soviet intercontinental ballistic missile (ICBM) and transfer it into a Mars Oasis carrier rocket. In late October 2001, Musk invited Jim Cantrell (who was engaged in international intelligence) and his college friend Adeo Ressi to fly with him to Moscow to try to buy an R-36 ballistic missile. Later, Michael Griffin also went to Moscow to join the negotiating team.

Russia offered a price of 8 million US dollars per rocket, while Musk had expected to get two for that price. Not only did the Russians reject his offer, but also ridiculed and insulted him. Musk realized that the Russians were not taking his business seriously, and wanted to squeeze as much money out of him as possible. He walked out and took a taxi straight to the airport.

On the flight back to the United States, Musk showed his friends a spreadsheet explaining the estimated cost of building and launching a rocket. Undaunted by his run-in with Russia, he said, 'I think we can build it ourselves.'

1.3 The Creation of SpaceX

In June 2002, Musk gave up on the idea of propagating the Mars program in public, and turned his hand to developing and launching rockets through SpaceX. The Mars Society and many others were disappointed by this move, as they had expected Musk to fail like so many other rocket entrepreneurs.

With the 1.5 billion US dollar acquisition of PayPal by eBay, Musk suddenly possessed the huge sum of 180 million US dollars, which allowed him to invest in SpaceX and other projects.

After four years of hard work and funding from the DARPA Falcon program, SpaceX finally launched its first rocket – the Falcon 1 – from its facility on March 24, 2006. However, the rocket fell back to Earth shortly after lifting off, and its first test launch was declared unsuccessful. In the same year, SpaceX successfully raised research funding from NASA for commercial cargo services for the International Space Station to develop a more powerful Falcon 9 rocket. Thanks to NASA's support for Falcon 9, Musk canceled the development of the smaller Falcon 5 rocket.

On March 21, 2007, the second Falcon 1 rocket failed to launch again. Unfortunately, due to SpaceX's successive failed launches and the fact that Tesla Electric Motors was running into trouble, Musk began to suffer a spate of unprecedented financial difficulties. He was forced to prioritize sending the Falcon 1 rocket into orbit in order to fill the financial gap.

On August 3, 2008, the third test launch of the Falcon 1 rocket failed. This was the worst day of Elon Musk's life. He only had enough money to pay for the one more launch, and if it failed, he would be bankrupt. Fortunately, on September 28, 2008, the fourth launch was successful.

However, the development of SpaceX still required a steady stream of funds from the outside world. On December 23, 2008, the company defeated many powerful competitors and was awarded a contract of 1.6 billion US dollars in COTS (Commercial Orbital Transportation Services) by NASA. Under the terms of the contract, SpaceX would have to use its own rocket and spacecraft systems to provide cargo replenishment services to the International Space Station. Only a few hours before the deadline for SpaceX and Tesla employees to pay their checks, Musk's dream of going to Mars finally looked as if it might come true.

Two years later in June 2010, Space X's first heavy rocket – the Falcon 9 – was successfully launched.

1.4 Musk's success: not a question of luck

Xu Kuangdi, former president of the Chinese Academy of Engineering, made a keynote report at the International High-End Forum on 'Development a Strategy for Mechanical and Transportation Engineering Technology 2035' held by Shanghai University. As a cutting-edge and game-changing step forward, true disruptive technology has two things in common. First, it is based on solid scientific principles – not a myth or fantasy, but an innovative application of scientific principles. Second, it is an interdisciplinary and cross-disciplinary integration of innovations – not a linear innovation in the fields of design, materials, and technology. Xu added that the most recent example of innovative success with disruptive technologies after Bill Gates and Steve jobs is Elon Musk.

How has Musk achieved so much success in a range of fields? It is not a question of luck. He has successfully integrated his innovative thinking and down-to-Earth management, and has honed his entrepreneurial philosophy almost to perfection.

First, he firmly grasped an opportunity. After graduating from college, his dream was to enter the three areas that he thought would have unlimited potential in the future: the Internet, clean energy, and aerospace. These three areas are closely related to the general trend of social sustainable development. He initially devoted himself to the Internet (PayPal), which was booming. He then established Tesla Motors, an electric car company, and invested in solar energy (Solar City), coinciding with the golden age of the low-carbon economy. Finally, he founded SpaceX, just at the time when the US government was offering support to private enterprises to enter the aerospace industry. Thanks to Musk's astute judgment, Tesla Motors received a loan of 465 million US dollars from the US Department of Energy. SpaceX received not only strong support from president Obama and the US government, but also technical support from NASA and billions of dollars in service contracts.

Second, Musk advocates innovative ways of thinking that are based on the basic laws of things. He said: 'Through most of our life, we get through by reasoning by analogy, which essentially means copying what other people do with slight variations. When you want to do something new, you have to apply the physics approach. Physics is really figuring out how to discover new things that are counter intuitive.' For example, when he was developing the Falcon rocket series, he started from the laws of physics and found that it was necessary to put substances weighing 'X' into orbit. It costs 'Y' by weight of fuel and 'Z' by weight of material, but for existing carrier rockets, 'Y' and 'Z' cost only 1% of the total. Therefore, he believed that there must be room for significant savings in the total cost of the rocket. Similarly, starting from the basic law of reliability, he followed the principles of simplicity and reliability in rocket design to ensure high reliability. From this, he was able to develop the world's lowest cost carrier rocket.

Thirdly, flat management architecture and powerful teams made it possible to improve efficiency and reduce costs. Musk believes that one of the main reasons for the high cost of space launches is the bureaucracy in traditional space companies. Therefore, SpaceX has adopted a highly efficient flat management mechanism, with no division of regular departments within the company. Employees in all fields can communicate freely in technical discussion, design, and development. This flat organizational structure not only ensures a close connection between R&D and production, but also saves management costs. SpaceX also reduces the grading of external subcontractors, which has accelerated decision-making and transportation to a certain extent, forming a rapid design and processing loop.

SpaceX's core staff come from NASA and traditional space companies, and are all members of the aerospace elite, with outgoing NASA administrator Michael Griffin at the head of Musk's think tank. SpaceX's ambitions have also attracted many young technicians who like to take risks. The company had only 160 employees in November 2005, when basic system development was completed. By July 2008, when the Falcon 1 rocket had completed its first flight, SpaceX had just over 500 employees. By 2010, the Falcon 9 rocket was being developed, the company had 1,100 employees. SpaceX has 25 employees and six task controllers at its control center on Kwajalein Atoll. The company's own Falcon 1 rocket team is made up of just 20 people. Employees are so skilled that the main goal is to reduce costs. With the expansion of its mission, SpaceX had more than 4,000 people by 2015. While the company is growing, it still has far fewer employees than the major US aerospace firms.

Fourthly, Musk is persistent in the pursuit of his beliefs. He has suffered a series of setbacks in developing rockets and electric cars, but he has never been discouraged. In 2008, the Falcon 1 rocket suffered its third failure and employees were in low spirits. Musk came out of the control room to speak to them, telling them that they had to pick themselves up and keep going. 'I think most of us would have followed him to the end,' recalls one of the employees who was present. In 2008, when the financial crisis hit the United States, Tesla Motors faced even more difficulties. Musk did not hesitate to put the last of his money into the company to make electric sports cars possible.

In addition to his own efforts, Elon Musk's success is largely due to his education and the innovative environment provided by society. It is not question of luck, which means that success like Musk's is now entirely possible in China. Among Chinese entrepreneurs, there are thousands whose wealth exceeds Musk's initial accumulation of more than 100 million US dollars, but few have been able to combine technological innovation with commercial operations. With the deepening of China's reform and consistent improvement of the innovation environment, a new generation of space entrepreneurs like Musk will certainly emerge.

2.   Jeff Bezos' Entrepreneurship

In China, less is known of Jeff Bezos, the founder of American start-up Blue Origin, and his New Glenn rocket. In fact, like Elon Musk, Bezos started from the Internet, and dreamed of space travel since childhood. The difference between them is that Musk likes to take risks and move fast, while Bezos prefers to play it safe.

2.1 Inherited ideals

Jeff Bezos was born in New Mexico in 1964. In 1986, he earned a Bachelor's Degree in Electrical Engineering and Computer Science from Princeton University, and then joined FITEL, a high-tech company in New York engaged in the development of computer systems. In 1988, Bezos joined Bankers Trust Co. on Wall Street, where he later served as Vice President. From 1990 to 1994, he helped to set up D. E. Shaw & Co., a fund trading management firm, and became Vice President in 1992. He founded Amazon.com in Seattle in July 1995, which became one of the most successful e-commerce companies in the world when its stock was listed in May 1997. Bezos named his company after the world's largest river, hoping that his company would echo its size and reach.

Bezos' grandfather, Lawrence Preston Gise, was a major in the US Navy during World War II. When he left the military, he went to work for the United States Atomic Energy Commission due to his strong interest in rocket technology. On retirement, he lived a simple life on his farm. Between the ages of four to 16, the young Bezos spent every summer on this farm, and witnessed his grandfather remarkable practical ability. He learned to clean the barn, castrate bullocks, and install water pipes. 'The first thing you learn in a place like that,' Bezos recalled, 'is to be on your own. My grandfather was one of my idols. If something breaks, let's fix it. To do something new, you have to be stubborn and focused.'

Bezos' grandfather was also interested in science fiction and space exploration, and encouraged the young man to pursue his dreams. At the age of 14, Bezos decided to become an astronaut or physicist, carrying out engineering and scientific experiments in his garage, such as changing the vacuum cleaner into a hydrofoil ship and making a solar cooker with an umbrella. As a teenager, Bezos developed the ambition and innovation that have shaped his career.

2.2 Amazon's Innovation

When Amazon was first establishment, it did not make much of a mark. In 1999, Thomas Friedman, author of *The World Is Flat*, wrote in a column in the New York Times, 'Amazon is destined to fail. Anyone can build something similar in their bedroom.' As it happened, Friedman's prediction was completely wrong.

Amazon has long surpassed the retail sector, including major players like Wal-Mart. In 2007, Amazon began selling digital audio, followed by on-demand video services, becoming a provider of entertainment content It then launched the e-book reader

Kindle, which opened up a brand new market. The Kindle is now Amazon's best-selling product, selling more e-books than paper books on Amazon.com.

When other tech companies were still dreaming of the future of Cloud computing, Amazon's AWS (Amazon Web Services) had more than 500 million US dollars in revenue by the end of 2010, and had 400,000 corporate customers, including the New York Times and NASDAQ stock exchange. It has also developed robots, which have greatly improved automation in its warehouses.

Amazon has been innovating and exploring since the start. It has transformed from the largest online bookstore into the largest integrated online retailer and then into a customer-centered enterprise. In recent years, with the development of Artificial Intelligence (AI), Amazon has begun to make a major push into physical stores. Its Amazon Go app allows people to go to supermarkets, pick things up, and leave without using a checkout. This AI supermarket retains the sense of experience and pleasure brought about by real business, as well as price comparison.

Amazon's share price has risen by more than 20% since 2017, and recently exceeded 900 million US dollars. This has made Bezos the second richest person in the world, surpassing Warren E. Buffett. Bezos' net worth has grown by 123% over the past five years. Judging by the growth rate of 2016, Bezos will soon catch up with Bill Gates, the world's richest man.

2.3 Blue Origin

As for Bezos' next goal on his entrepreneurial path, he turned his attention to space. He founded a company called Blue Origin in 2000, and spent huge sums of money to buy thousands of acres of land in Texas for rocket and spacecraft flight tests. On April 5, 2017, Bezos announced at the 33rd Space Symposium that he had decided to sell about 1 billion US dollars of Amazon's stock each year to support Blue Origin's ambitious space goals. The company's ongoing work includes the construction of the New Shepard suborbital space tourism rocket, the testing of the BE-4 large thrust liquid oxygen/methane fuel engine, the development of the New Glenn reusable launch vehicle, and the detailed design of the Blue Moon unmanned lunar lander. The company's existing staff has expanded from the initial 10 to more than 1,000, and will recruit hundreds more people in 2017. Bezos firmly believes that Blue Origin will be able to stand on its own as a profitable company with a strong business.

3.   The Opportunities of Commercial Manned Space Missions

3.1 The 'Magic' Dragon Spacecraft

At 03:44 local time on May 22, 2012, the Falcon 9 rocket carrying the Dragon spacecraft was launched from Cape Canaveral Air Force Base in Florida. Launched by SpaceX, the Dragon was carrying about half a ton of supplies to the International Space Station. In the early hours of May 25, two astronauts on the Space Station manipulated a robotic arm to capture the Dragon spacecraft. On May 31, about six days after the Dragon berthed, it returned to Earth with more than 600 kilograms of payload.

It was the world's first commercial spacecraft to berth on the International Space Station. In a statement on the Space Review website, Barack Obama's chief science adviser John Holdren said: 'Every launch into space is a thrilling event, but this one is especially exciting because it represents the potential of a new era in American space travel – "The Big Test". However, many perceive the mission as something of a test of the concept of commercial transport to the ISS, not just for cargo but also for crew, particularly as Congress debates the future of the commercial crew program.'

The Dragon spacecraft is about 6.1 meters high and 3.7 meters in diameter, and comes in both cargo and manned forms. The flight test was the cargo type, which can transport up to six tons of cargo and bring back up to three tons. SpaceX plans to launch a manned Dragon spacecraft in 2018, which is capable of carrying seven people. The Dragon is not like generic spacecraft that can only be used once. It can be reused about 10 times, so it can greatly reduce the transportation cost between orbit and Earth. On June 3, 2017, SpaceX successfully launched the first reusable Dragon spacecraft using its Falcon 9 rocket, and plans to combine manned and cargo spacecraft in the future.

When SpaceX was founded in 2002, many aerospace experts believed that Elon Musk could not succeed, which is why Musk named the spacecraft Dragon after the nursery rhyme song 'Puff the Magic Dragon':

> *Puff the magic dragon lived by the sea*
> *and frolicked in the autumn mist in a land called Honah Lee.*
> *Little Jackie Paper loved that rascal Puff,*
> *and brought him strings and sealing wax and other fancy stuff. Oh!*
> *Together they would travel on a boat with billowed sail.*
> *Jackie kept a lookout perched on Puff's gigantic tail.*
> *Noble kings and princes would bow whene'er they came.*
> *Pirate ships would lower their flag when Puff roared out his name. Oh!*

A dragon lives forever but not so little boys;
Painted wings and giant rings make way for other toys.
One grey night it happened: Jackie Paper came no more,
and Puff that mighty dragon, he ceased his fearless roar.
His head was bent in sorrow, green scales fell like rain,
Puff no longer went to play along the cherry lane.
Without his life-long friend, Puff could not be brave,
So Puff that mighty dragon sadly slipped into his cave. Oh!

Elon Musk had loved this nursery rhyme since childhood. He named his spacecraft after the eponymous Dragon to stimulate his team's imagination and pursuit of simplicity, which is an important principle by which SpaceX improves reliability and safety.

The history of world aviation shows that shortly after the Wright brothers achieved first sustained powered human flight in December 1903, some people attempted to use aircraft for transportation and postal services. However, it wasn't until December 1935, when the all-metal airplane DC-3 made its maiden flight, that airlines became commercial. Manned spacecraft technology is far more complex than that of passenger aircraft, and the road to commercial manned space travel is expected to be longer and more complex. Therefore, technological progress and scientific management are required.

Thanks to his disruptive innovations in many major areas, Elon Musk topped Fortune magazine's list of the Top People in Business for 2013. The magazine also compared his contributions to those of Steve Jobs.

3.2 The US: speeding up the commercialization of space

Early in the morning of April 22, 2013, Beijing time, the US Orbital Science Corporation successfully launched an Antares launch vehicle. Since the purpose of this launch was only for testing, it did not carry Orbital Science's Cygnus spacecraft, but bore a heavy load of 3.8 tons, equal to the weight of the spacecraft. The mission also successfully launched three NASA phone satellites.

The launch of Antares proved that another American private company possessed the capabilities for space launches. This laid the foundation for the Orbital Science corporation to launch the Cygnus spacecraft to the International Space Station. As NASA shifts its focus to more challenging missions, such as preparing to send people into deep space, private companies will take the baton in terms of putting astronauts and related cargo into low-Earth orbit. Former NASA director Charles Bolden said that

NASA hopes to rely on US-based companies to deliver supplies and astronauts to the International Space Station. Therefore, today's success is an important milestone.

Orbital Science was founded in 1982 by several American aerospace engineers. Over the past 20 years, it has made remarkable achievements in the development of air-launched rockets, small satellite communication constellations, and commercial remote-sensing satellites. Its Antares 130 is a two-stage rocket that is 40.5 meters long and 3.9 meters in diameter, with a maximum launch weight of 24 tons and a maximum payload of 6,350 kilograms to reach low-Earth orbit. The first stage is powered by two AJ26-62 liquid rocket engines manufactured by US Aerojet and fueled by liquid oxygen/kerosene. The prototype engine is NK-33 – an engine developed by the Soviet Union for a failed manned N-1 rocket landing on the Moon. After the collapse of the Soviet Union in the 1990s, Aerojet imported the engine into the United States and made some improvements. In 2007, Orbital Science decided to put the engine on the Antares rocket, then known as Taurus 2. The second stage of the Antares rocket is a solid rocket engine.

NASA launched the Commercial Orbital Transport Services (COTS) project in 2008, aiming to develop commercial resupply services for the International Space Station. After a round of bidding, NASA finally chose two innovative private companies to deliver cargo to the ISS. NASA signed a 1.6 billion US dollar contract with SpaceX to deliver 12 cargo flights to the ISS using its Falcon 9 rocket and Dragon spacecraft. Before the launch of the Antares rocket, the Dragon spacecraft had completed two delivery missions. Within the 1.9 billion US dollar contract, Orbital Science's spacecraft will deliver 20 tons of cargo to the ISS eight times between 2013 and 2016.

Commercial companies in the United States have been responsible for a number of past successes in the field of manned space travel. NASA itself has never been involved in building rockets or spacecraft, leaving the actual design and construction to private contractors such as Boeing and Lockheed Martin. The difference between NASA's current COTS project and the previous one is the way that funds are used. In the past, NASA has offered to pay for development and ensure profit, regardless of how the company performs. COTS pays a fixed amount of funds to these companies after bidding. If the development costs exceed the expenditure, the bidding company will bear the responsibility. Therefore, private companies bear the main risks.

After the success of COTS, NASA launched a second round of Commercial Resupply Services (CRS-2) to commercial companies competing in the procurement of orbital transportation services for logistical support to the ISS. In January 2016, NASA announced that the winners of the second cargo resupply services were SpaceX, Orbital ATK (a new company formed in 2014 after the merger of Orbital Science and ATK Technologies Systems; Orbital ATK has now merged in Northrop Grumman) and

Sierra Nevada Corp., which will be awarded contracts starting from 2019 through 2024. Respectively, they will use Dragon spacecraft, Cygnus spacecraft, and Dream Chaser spacecraft to provide cargo delivery for the ISS. The first flight will be carried by the Dream Chaser spacecraft in 2019, propelled by the Atlas V launch vehicle.

3.3 The advent of commercial manned spacecraft

In 2009, the Obama administration asked NASA to identify some competitive companies in the United States to develop manned spacecraft and rebuild the nation's space delivery capacity. That same year, NASA launched the Commercial Crew Development (CCDev) Program to support future crew transportation on the ISS.

The first round of CCDev was launched in February 2010 with the aim of completing the development of system concepts and key technologies. NASA awarded a total of 50 million US dollars in contracts to four companies through the Space Act Agreements. Among them, Blue Origin is responsible for the development of the new Pusher Escape System and Composite Pressure vessel; Boeing is responsible for the development of Crew Space Transportation (CST) in the form of 100 Starliner spacecraft; Paragon Space Development Corporation is responsible for the development of the Commercial Crew Transport Air Revitalization System (CCT-ARS) and the United Launch Alliance; and LLC is responsible for the development of the Emergency Detection System (EDS) for the Evolved Expendable Launch Vehicle (EELV) manned launch system.

The second round of the CCDev program was launched in October 2010, and focused on the design and maturity of spacecraft and launch vehicles in order to accelerate the realization of Commercial Crew Transportation System (CTS) technology. NASA awarded 269.3 million US dollars to Boeing, Sierra Nevada Corporation, SpaceX, and Blue Origin. Boeing went on to develop the CST-100 spacecraft, while Sierra Nevada developed the Dream Chasers spacecraft, and SpaceX developed the Dragon 2 integrated launch abort system.

The third round of CCDev was launched in February 2012, and is also known as Commercial Crew Integrated Capability (CCiCap). NASA wanted a complete list of proposals for CTS technology, including spacecraft, launch vehicles, ground operation, and mission control, to help it achieve safe, reliable, and economical manned access to space. NASA awarded contracts to SpaceX, Boeing, and Sierra Nevada in August 2012 to continue developing and testing manned transport systems.

On November 19, 2013, NASA released a proposal of final requirements for the new Commercial Crew Transportation Capacity (CCtCap) contract to ensure that the Crew Transportation System (CTS) built by commercial companies meets NASA's safety requirements for transporting astronauts. In September 2014, NASA eventually sold a

6.8 billion US dollar CCtCap contract for the International Space Station (ISS) manned transport mission by SpaceX (2.6 billion US dollars) and Boeing (4.2 billion US dollars). In May and November 2015, Boeing and SpaceX won the first NASA commercial crew capacity (CCtCap) contracts respectively.

NASA has stated that the aircraft of the two companies must pass safety tests before manned flight. After passing flight certification, each company will undertake two to six manned transportation missions to the ISS from 2017. Boeing's contract stipulates that it will build three CST-100 spacecraft at the Kennedy Space Center in Florida, each capable of carrying seven passengers. On September 4, 2015, Boeing officially named the CST-100 manned spaceship 'Starliner'. Its shape is similar to the Apollo and Orion spacecraft, and is between them in terms of volume. Its maiden launch is slated for the end of 2018.

NASA's choice of commercial manned spacecraft means that no space-related decision in any country can be a purely technical choice; it is inseparable from its domestic and international political environment. As a core member of the military complex, Boeing has a major influence on Capitol Hill. When NASA awarded Lockheed Martin, Boeing's biggest rival the contract for the Orion spacecraft, it compensated by also supporting Boeing's CTS-100 spacecraft, which is very similar in configuration. Sierra Nevada Corporation's Dream Chaser spacecraft had an accident when it landed after its first flight test in 2013, and progress was obviously behind schedule. In addition, after Boeing's successful development of the X-37B, it also proposed the manned X-37C as a spare, so NASA no longer needs to worry about the loss of space shuttle technology. Finally, the company relied on Boeing's Atlas V launch vehicle, while SpaceX's Dragon 2 manned spacecraft was launched with its own Falcon 9 heavy rocket.

SpaceX did not encounter much trouble in this bid. The end result was that it took a small share while Boeing took a large share. The reason for this was first of all, if SpaceX is eliminated and only Boeing is left, the big contract from NASA will always be awarded to either Boeing or Lockheed Martin. There is no difference. This is contrary to Obama's original intention to encourage private companies to enter the space field. Secondly, the total cost of CTS-100 is obviously higher than that of the Dragon 2 spacecraft. Only by retaining SpaceX can the total cost of the CTS-100 spacecraft be reduced.

On March 2, 2019, SpaceX successfully launched the first manned Dragon spacecraft with the Falcon 9 carrier rocket at the Kennedy Space Center in Florida. This was an unpiloted test flight. After 27 hours, the manned version of the Dragon spacecraft successfully docked with the International Space Station. It is estimated that a manned flight of the Dragon will be achieved in July 2019.

3.4 The Antares Rocket Explosion

At 18:22 EDT on October 28, 2014, Orbital ATK launched an Antares 130 carrier rocket, attempting to deliver 2,300 kg of supplies to the ISS on the Cygnus cargo spacecraft. It exploded six seconds after lift-off, destroying the rocket and cargo satellite.

Antares-130 is a two-stage solid-liquid hybrid launch vehicle. The difference between this rocket and the previous two is that the second-stage solid rocket engine is slightly different. Judging from the flight timing sequence, the rocket explosion happened just after the first-stage engine was ignited. Such a strong explosion is likely to be caused by a failure of the first AJ26 engine. In May 2014, an engine exploded during a ground test, caused by a stress fatigue fracture.

On October 29, 2015, an NASA Independent Review Team (IRT) report was released on the explosion of the Antares-130 rocket. The report concluded that after 15 seconds of ignition, the E15 Liquid Oxygen (LO2) turbo-pump exploded inside one of the two AJ-26 engines on the first stage of the rocket. The report also stated that the cause of explosion was a loss of rotor radial positioning resulting in contact and frictional rubbing between rotating and stationary components within the Engine LO2 turbo-pump Hydraulic Balance Assembly (HBA) seal package. This led to ignition and a fire involving LO2 within the turbo-pump HBA. However, the report did not identify the root cause of the turbo-pump failure. Orbital ATK's report gives a technical reason for the accident: a manufacturing defect in the assembly of the turbo-pump.

Finally, Orbital ATK abandoned the AJ-26 engine that had led to the Antares rocket accident and switched to the Russian RD-181 engine. On October 17, 2016, the new Antares rocket was successfully launched, delivering cargo supplies to the ISS.

The Antares rocket explosion delayed the Orbital ATK's implementation of NASA's COTS mission for two years, but the accident did not change US policy on commercializing manned space. In addition, the incident impelled private aerospace enterprises to vastly improve the reliability and safety of their rockets. It also inspired the US aerospace industry to alter the nation's passivity in terms of relying on Russian technology for liquid oxygen/kerosene rocket engines.

Chinese American NASA astronaut Leroy Chiao said to CNN after the accident: 'Without a doubt, critics will question why we are entrusting cargo deliveries and future crew exchanges to commercial companies. The answer is simple: It is a logical evolution of technology and commercialization, following the same path as the development of the airplane and commercial air transportation. This mishap was painful, but it is only a speed bump on the way to the commercialization of space travel. We've been launching astronauts into space for over 50 years. The technologies mature. It's a matter of seeing if we can create a commercial environment for these companies to make a profit, and then let NASA buy those services, rather than have to run it itself.'

3.5 The commercial space station: not a myth

In early 2013, NASA announced that it had awarded a 17.8 million US dollar contract for the construction of the Bigelow Expandable Activity Module (BEAM) for the ISS. Bigelow Aerospace is a private company that builds cheap commercial space stations for low-Earth orbit (LEO). 'Today we're demonstrating progress on a technology that will advance important long-duration manned space travel goals,' remarked Lori Garver, NASA's Deputy Administrator, at the signing. 'NASA's partnership with Bigelow opens a new chapter in our continuing work to bring the innovation of industry to space, heralding cutting-edge technology that can allow humans to thrive safely and affordably in space.'

Bigelow Aerospace was not the first company to create inflatable modules. TransHab was originally designed in 1997 by William Schneider, a Senior Engineer at NASA. The design was also considered for use as a habitation module for the ISS, which was being developed by Boeing at the time. However, in 2000, before it could produce a real lift-off product, the project was inexplicably canceled.

Bigelow Aerospace was founded by the American billionaire Robert Bigelow. Born in the renowned casino city of Las Vegas, he had dreamed of space since childhood, but his career in the first half of his life had nothing to do with it. He studied for an MBA at the University of Arizona, and since his father was a real estate dealer, he went into the real estate hotel business after graduating in 1967. He founded a hotel chain – the Budget Suites of America – in 1988. By 1999, his assets had reached more than 1.5 billion US dollars. In the same year, Bigelow founded Bigelow Aerospace and announced that it would spend 500 million US dollars to build a space hotel.

Bigelow later signed the Space Act Agreement (SAA) with NASA. According to this agreement, Bigelow not only had the rights to develop the TransHab project, but would also be able to invite NASA engineers who had participated in the project to work for the company. Bigelow met William Schneider, who had retired from NASA and was now teaching at Texas Agricultural and Industrial University, and the pair hit it off.

Bigelow Aerospace's space hotel test module, which is made of inflatable soft-shell modules, has the advantage of being smaller, lighter, and cheaper than traditional metal capsules. After years of planning and development, Bigelow Aerospace built its first space habitat test module, Genesis I, in October 2005. It has an internal volume of 11.5 cubic meters, and was launched by a Russian rocket on July 12, 2006. The larger Genesis

II was also launched by a Russian rocket on June 28, 2007. Genesis II was only 1.6 meters in diameter when it was launched, but can expand to 2.54 meters in diameter once in orbit. Compared with the Genesis I, Genesis II has made great technical improvements including replacing the single-tank inflation system with multiple tanks, and improving the heat protection, thermal management, and control systems. Since Genesis II was designed to last 12 years, it is still in orbit. Both inflatable capsules carry scientific, commercial, and entertainment payloads.

On April 10, 2016, SpaceX's Dragon spacecraft arrived at the International Space Station. Among the 3.18 tons of cargo delivered this time, the most notable was undoubtedly the Bigelow Expandable Activity Module (BEAM). As the first inflatable module on the ISS, the successful berthing of BEAM is of great significance. If successful, it could become the basic form of a future space base for humans. It weighs 1.4 tons, and is made of aluminum and special foldable fabrics. It will be compressed during flight to form a 'parcel' with a length of 2.4 meters and a diameter of 2.36 meters. After berthing with the ISS, the length and diameter of BEAM will increase to 3.7 meters and 3.2 meters respectively, and the internal space will expand from 3.6 cubic meters to 16 cubic meters, which is equivalent to the size of a small bedroom. After berthing, BEAM inflated successfully. According to the plan, it will stay on the ISS for two years. During this period, astronauts will enter it several times a year to install instruments and equipment, collect data, and assess its status, but will not live in the inflatable compartment.

Since NASA's future efforts will focus on manned missions to Mars, plans for a low-Earth orbit space station and a return to the Moon may be carried out by private space companies. In addition to testing inflatable modules on the ISS, Bigelow Aerospace plans to build commercial space stations that can operate independently. In 2020, the company will use ULA's Atlas V rocket to launch a B330 inflatable module that is larger than BEAM into space. As its name suggests, each B330 module has 330 cubic meters of pressurized space after inflatable expansion. Currently, the internal pressurized space of the ISS is 916 cubic meters. A B330 can provide one third of the current capacity of the ISS when it is deployed in space. Moreover, the total weight of the three B330 modules is only about 60 tons, while the mass of the ISS is as high as 420 tons. This can slash the launch cost by a significant amount.

NASA believes that this scalable architecture may be a key technology to help humans reach Mars. Because the inflatable space modules are lower in cost and easier to control in weight and volume, they are especially suitable for deeper space missions that take longer. In such missions, astronauts will need more space to store supplies, work, live, and entertain themselves.

4. Space Tourism: Not Just an Impossible Dream

More than half a century has passed since Soviet astronaut Yuri Gagarin entered space in 1961. For many years thereafter, manned space travel was dominated by governments and states, but this is no longer the case. On April 28, 2001, an American businessman named Dennis Tito became the first civilian to enter the ISS, beginning the era of commercial space tourism.

The only way to enter space as a tourist is to take the Russian Soyuz spacecraft to the ISS, 400 kilometers away from the Earth, and experience seven to 10 days of astronaut life, but this will cost tens of millions of dollars. Between 2001 and 2008, Space Adventures sent seven tourists – seven men and one woman – to the ISS on the Soyuz, one of whom had been before. Their eight tickets totaled 250 million US dollars. However, after 2008, the Soyuz was once again occupied by professional astronauts.

4.1 The prize-winning SpaceShipOne

In order to accelerate the development of reusable space flight technology, the XPrize Foundation in St. Louis US established the 10 million US dollar Ansari X Prize in 1997, requiring aircraft to fly twice to space beyond 100 kilometers within a maximum of two weeks. The winner was SpaceShipOne (SS1), developed by Scaled Composites. Its first flight test was conducted in the Mojave Desert on June 21, 2004. The carrier aircraft, called White Knight, launched the SS1 at an altitude of 15 kilometers. When the rocket engine ignited, the SS1 was sped up to about Mach 3, and then climbed vertically to almost 50 kilometers. After the shutdown of the rocket engine, the SS1 rose vertically into space, reaching about 103 kilometers above the ground. Weighing less than three tons, the SS1 stayed at this altitude for several minutes. During this period, the crew experienced a feeling of weightlessness until the vessel returned to an altitude of about 60 kilometers. Then, the SS1 began its descent, with the back half of its wings and two tail fins rotating upward until they were at right angles to the ship to help slow it down. When the SS1 returned to low altitude, the tail rotated back to its original position, and the pilot glided without power to land like a normal plane in the Mojave Desert.

White Knight took off again at about 07:00 on September 29, 2004 carrying the SS1. However, shortly after the spacecraft hooked off from the White Knight, it was caught at an unconventional attitude. The SS1 underwent an abnormal rotation as it crossed the edge of the atmosphere, and flipped more than 30 times during its descent. Fortunately, after exhausting its rocket fuel, the spacecraft returned to normal. At around 08:00 local time on October 4, the SS1 officially entered space and landed safely.

SpaceShipOne has many innovations in its design and integration, such as the rotating of the rear wing on return, as mentioned above. However, overall, its distinctive features are the use of aircraft to take off, and the employment of solid-liquid hybrid rocket engines. First of all, compared with direct lift-off, it not only saves at least half of the propellant, but also easily achieves horizontal take-off and landing at ordinary airports. Second, the use of a solid-liquid hybrid rocket engine makes the engine safer in operation, and can also greatly reduce costs.

4.2 Difficulties for the SpaceShipTwo

After the SpaceShipOne won its award, British billionaire Richard Branson announced that he would be setting up a space tourism venture company called Virgin Galactic, to invest in the development of SpaceShipTwo (SS2). Its design is similar to that of the SS1, but it has a more spacious cockpit and a better view from the porthole (up to half a meter wide). The wingspan of the SS2 is 8.2 meters, and it is 18 meters long and 4.6 meters high. The manned cabin is 3.7 meters long and 2.3 meters in diameter, and it can carry two pilots and six passengers. The SS2 is carried to a height of 15 kilometers by the White Knight 2, with a wingspan of 34–38 meters. After separating from the carrier aircraft, its single solid-liquid hybrid rocket engine ignites and propels the aircraft to fly at supersonic speed, with a maximum speed of 4,000 km/h. After 70 seconds of operation the engine shuts down, and inertia allows the aircraft to reach a height of 150 kilometers. Virgin Galactic announced that each suborbital flight will cost 200,000 US dollars. At present, more than 700 people around the world have booked tickets.

At 12:20 on October 31, 2014, the SpaceShipTwo (SS2) took off from the Mojave Air and Space Port carried by White Knight 2. Fifty minutes later, it separated from the carrier aircraft. Shortly after, the spacecraft violently broke apart and crashed 40 kilometers north of where it had taken off. The accident claimed the life of co-pilot Michael Alsbury and caused serious injury to pilot Peter Siebold.

The preliminary investigation into the SpaceShipTwo incident showed that the fuel tank and engine were intact at the time of the accident, but the feathering system (the craft's air-braking descent device in re-entry) was found to be out of order. The system should have been unlocked when the ascending flight speed reached Mach 1.4, but it was unlocked at Mach 1.0, causing the SS2 to disintegrate under heavy aerodynamic loads. The accident investigation by the National Transportation Safety Administration (NTSB) showed that the co-pilot unlocked the feathering system at the wrong moment, which led to the disintegration of the SS2 in mid-air. The report also clarified that Alsbury's simulation training had not properly enabled him to deal with extreme turbulence and overloading during the flight.

Undoubtedly, the weak point of this innovative approach of launching orbital aircraft with carrier aircraft is still its safety. The day after the crash, Richard Branson vowed at the scene to move on, but safety was the key. He said: 'We do understand the risk involved, and we're not going to push on blindly. To do so would be an insult to all of those affected by this tragedy. We're going to learn from what went wrong, discover how we can improve safety and performance, and then move forward together. '

On February 19, 2016, Virgin Galactic unveiled its new SpaceShipTwo. The name VSS Unity was chosen by renowned British physicist Stephen Hawking, and it completed its first testing flight on September 8, 2016. During the test, the spacecraft was mounted on the White Knight 2 carrier craft, and did not fly alone. For the purposes of the test, White Knight 2 played the role of a flying wind tunnel, carrying out real airflow on the spacecraft and testing its performance in extremely cold conditions at an altitude of 15 km. The test flight was completed by four pilots, two at the controls of the White Knight 2 aircraft and two in the spacecraft.

On December 14, 2018, the SpaceShipTwo conducted a flight test, carrying two passengers to an altitude of 82.7 kilometers and crossing the threshold of space. On February 3, 2019, another flight test was carried out, carrying three passengers to an altitude of 89.9 km.

5. Embracing Aerospace Entrepreneurship

The emerging space economy is changing all aspects of human life on Earth. Its development and the growing maturity of basic space technology have provided many exciting opportunities for entrepreneurs.

5.1 Skybox

At the end of 2007, Google announced the Google Lunar X Prize of 20 million US dollars. The competition required teams to send their spacecraft to the surface of the Moon, explore its surroundings, take high-definition photos and videos, and transmit them back to Earth. This competition has inspired a new generation of entrepreneurs.

In 2004, a Stanford PhD student named Dan Berkenstock heard about the Google contest. Along with several of his fellow students, and with support of investors, he decided to enter. They spent a year on the project, but with the financial crisis of 2008, almost all investment was halted.

At this point, Berkenstock knew that his project needed only one more push to revolutionize the aerospace industry. He also knew that he had a very talented group of friends around him, so he formed a team with three other Stanford students. They were John Fenwick (a US air force veteran), Julian Man (founder of the Aerospace Development Company), and Ching-Yu Hu (a former analyst at JP Morgan). At the beginning of 2009, Skybox was finally established as a four-person company. The initial plan was to launch smaller, cheaper satellites. Skybox started with a 10-centimeter cube satellite called CubeSats developed by Stanford University in 1999, but it was too small and limited in function. Skybox then began to develop medium-sized satellites weighing 100 kilograms, which could carry larger-diameter instruments and support more powerful computing capacity. The satellite's Earth observation resolution was one meter, which represented the best compromise between performance and cost. Skybox reportedly spent about 5 million US dollars to build each satellite.

The biggest hurdle Skybox has faced in its development has been financing. At the time, venture capital firms and the entire American aerospace industry believed that this small satellite company was a 'joke' and a 'naïve daydream'. However, Bessemer Venture Partners was so impressed by Berkenstock's entrepreneurial spirit that it gave Skybox 18 million US dollars in start-up funding in 2010.

On November 21, 2013 – the fifth year since Skybox was founded – the company successfully launched its first SkySat1 satellite on a Russian Dnieper rocket, with a total of 32 satellites planned for the future. The satellite operates in the solar synchronous orbit of the Earth, and its biggest technological innovation is the replacement of conventional scanning technology with a rapidly-updated complementary metal oxide semiconductor (CMOS) video imaging sensor, which has a panchromatic image resolution of 0.9 meters. On July 8, 2014, the company launched the SkySat2 satellite on a Russian Soyuz rocket and a constellation of 15 to 24 satellites is planned for the future. On December 11, 2013, Skybox released the first Earth observation image taken by SkySat1, and on December 27, 2013, it released the first Earth observation video taken by SkySat1. On July 10, 2014, the day after the launch of SkySat2, the company released the first Earth observation image taken by the satellite. In February 2014, Skybox signed a contract with Loral Space Systems to provide 13 Earth observation satellites, each weighing 120 kilograms.

In fact, more than 20 years ago, companies emerged in the United States offering high-resolution commercial Earth imaging products and services, such as DigitalGlobe. However, satellites developed by these companies (such as QuickBird) weigh upwards of 1,018 kg and costs hundreds of millions of dollars to build, making it difficult to update images quickly. Skybox's greatest achievement is that it has built much cheaper near-Earth satellites. Although the images are not as detailed as QuickBird's, they can be updated quickly, and video can also be provided. 'The Earth is a big database,' Berkenstock once said. 'You have to keep updating the data to capture changes in the global economy.'

In its time, Skybox was so unique that many large public companies in the United States wanted to buy it. In June 2014, Google acquired it for 500 million US dollars, which will allow it to use Skybox's satellite imagery to update Google maps and promote the Internet in remote areas, and integrate satellite data with its Knowledge Graph. For Berkenstock, this acquisition was the greatest prize he had ever won.

On May 8, 2016, the new Skybox headquarters was renamed Terra Bella. In 2017, Planet Labs acquired Terra Bella from Google.

As an entrepreneurial model focusing on technological innovation and financing on a small scale, Skybox is a good example for young people who want to start their own businesses in the aerospace industry after leaving university in China.

5.2 Copenhagen Suborbitals

Developing rockets to send people into space is expensive, risky, and traditionally an exclusive affair of the state. Founded by Elon Musk, SpaceX has embarked on a path of innovation, as the USA encourages private enterprise into manned space. However, the story of Copenhagen Suborbitals offers a completely different perspective.

Copenhagen Suborbitals (CS), led by Kristian von Bengtson and Peter Madsen, is a non-profit organization founded in Copenhagen in 2008, funded entirely by donations. Their mission is to build a rocket independently and carry a person into sub-orbit.

Peter Madsen was an inventor and entrepreneur who was responsible for developing rocket engines. He led the design and building of three submarines before starting Copenhagen Suborbitals (CS). Kristian von Bengtson earned his master's degree from the International Space University, where he was responsible for developing manned spacecraft. He also participated in the design of the Lunar Rovers and the preparation of NASA's Manned Integrated Design Manual.

Since its inception, the Copenhagen Suborbitals team has grown from around 20 people to more than 40. They initially raised about 71,300 US dollars, and spent a year and a half building a HEAT-1X rocket (HEAT is an abbreviation for Hybrid Exo Atmospheric Transporter). In September 2010, with the help of the Danish Navy, the team launched a 1.6-ton rocket from a submarine in the Baltic Sea. Unfortunately, the first launch failed because a fuse was blown. In June 2011, the HEAT-1X rocket was launched again from a mobile launch platform in the Baltic Sea carried by a standard test dummy, but the launch ended 21 seconds later when it reached an altitude of 2.8 km and began to deviate from orbit. On June 23, 2013, the team successfully launched the SAPPHIRE-1 rocket with simple control, reaching an altitude of 8.5 km.

Copenhagen Suborbitals has conducted extensive ground tests for rocket engines. HEAT-1X uses a solid-liquid hybrid rocket engine; the solid fuel is epoxy resin, and the oxidant is nitrous oxide. In the ground test, the team also made studies and tests using liquid oxygen to push paraffin wax, but the thrust was not large enough. Later, they tried using polyurethane instead of paraffin wax, but combustion instability occurred.

If the performance of rocket continues to improve, Copenhagen Suborbitals will launch a manned spaceship (called the Tycho Brahe 1 after the 16th-century Danish astronomer who was known for his extremely accurate astronomical observations before telescopes were invented) with a diameter of two meters as soon as possible, and Madsen will be the first pilot to go into space.

The team presumes that when the rocket enters sub-orbit, the spaceship will separate from the rocket and move in parabolic motion, then re-enter the Earth's atmosphere. The spaceship will then deploy a brake parachute to slow its flight speed, and gently descend to the water surface using three parachutes. The spaceship is small and can only accommodate one person, who must maintain a straight posture. The astronauts are unable to move freely, but their two arms are free to move and can control the spaceship. On top of the spaceship is a piece of plexiglass that allows astronauts to enjoy the view. In August 2012, the CS team tested the spaceship's Launch Escape System at sea. The craft did not rise high enough to open the parachute, so there was some damage when the ship fell into the sea.

Copenhagen Suborbitals is funded entirely by sponsorship. On October 5, 2010, a group of space enthusiasts set up the Copenhagen Suborbital Support (CSS) group. The main task of the organization is to support CS's economic, moral, and practical mission, and members make contributions at a fixed time. Within two days of its founding, the group had reached 100 members. It reached 500 members on November 15, 2011, and 1,000 members in December 2013.

Copenhagen Suborbitals has not escaped internal strife. On February 23, 2014, the board announced that Bengtson had left the project to focus on developing a manned Mars spacecraft called Mars I. On July 10, 2014, Madsen also announced his departure from the project, stating that he was preparing for another project that had exactly the same goal as the CS: sending a person into space.

At present, under the leadership of the Executive Committee, Copenhagen Suborbitals continues to develop. On July 23, 2016, it successfully launched the NEXO I rocket and recovered it at sea. The company is now preparing to launch the NEXO II rocket.

While the technological innovation and operation model of Copenhagen Suborbitals may be refreshing, the infighting within the cooperative organization also provides profound lessons.

5.3 From volunteer to entrepreneur

As a guest at the 2014 Tencent WE conference in November 2014, Jane Poynter spoke enthusiastically about WorldView, a company she had recently founded with her husband, Taber MacCallum. Using helium balloons and parachutes, the company was pioneering low-cost space travel for the general public, with passengers traveling 30km from Earth to the edge of space in a capsule for only 75,000 US dollars. In her speech, Poynter also revealed that Tencent had invested in the company. Mark Kelly, a former astronaut on the Space Shuttle Endeavour, joined the company in December 2013 as director of the flight crew.

According to Poynter, the flight time of a space capsule trip is about five hours. Before the dawn, passengers put on their flight suits, and then after a lift-off process of 1.5 hours, they climb 100,000 feet in the air and float for one or two hours as they enjoy breathtaking views of the Earth. It takes one hour to complete the landing process. Meals, toilets, and even a bar are provided.

Americans know Jane Poynter and Taber MacCallum because they participated in the renowned Biosphere II experiment. Related to Biosphere I, Biosphere II aims to test whether humans can survive outside of the Earth, creating a setting that completely mimics the Earth's ecosystem in the hope of developing a completely new living space on other planets. In the early 1990s, eight people, including Poynter and her husband, lived for 21 months in an area the size of two and a half football fields in an artificially enclosed ecosystem near Tucson, Arizona. In the Biosphere II experiment, scientists set up a number of miniature ecological types in this closed space to simulate the Earth's environment, such as rainforests, deserts, oceans, and swamps. All food came from their own cultivation, and the air and drinking water was drawn from internal circulation. By the sixteenth month of the study, the participants had developed severe fatigue and sleep apnea. A significant drop in oxygen levels was discovered inside Biosphere II. When one of the participants fell into a state of confusion, the area had to be re-oxygenated. The failure of this experiment was partly due to the inability to realize a virtuous cycle of oxygen, and partly due to the maladjustment of the relationship between biological species.

Poynter and MacCallum had been fascinated by space since childhood. Poynter grew up in England and came to the USA after high school. Her interest in space came from the science fiction of authors such as Isaac Asimov, whose work she read as a teenager. MacCallum's father was an astronomer, and his grandfather helped the Wright brothers build propellers. After graduating from high school, Poynter and MacCallum both joined a challenging training program in order to participate in Biosphere II. They went to a farm in the Australian outback, and spent time on an ocean research ship that sailed around the world. It was during this period that the couple met and fell in love. They married nine months after leaving Biosphere II.

Participation in Biosphere II delayed their chances of higher education, but the experience allowed them to acquire a lot of knowledge that could not be learned in the classroom. It also inspired them to become entrepreneurs, leading them to set up the Paragon Space Development Corporation, which aimed to find the most reliable two-way trip to Mars. Twenty years later, the company now employs around 70 engineers and scientists, and it is still expanding in scale. In December 2012, in collaboration with Golden Spike Space-Tourism, Perfect Space Development created a set of technologies related to spacesuits, temperature control, and life support systems for future commercial lunar travel.

Poynter and MacCallum also participated in the Inspiration Mars program proposed by Dennis Tito. The project aimed to launch a spacecraft in 2021, with a round trip time of 580 days. Program staff believed that, like Mars One, it would be difficult to succeed on a one-way trip only by private forces, so they hoped to cooperate with NASA. In a later NASA hearing, Tito said that he needed about 1 billion US dollars of government funding for the development of space launch systems in the following four to five years. Obviously, NASA did not agree, so the program had to be suspended temporarily.

Poynter and MacCallum believe that whether the Inspiration Mars mission is successfully or not, and whether they can go to Mars in person, their efforts will inspire public enthusiasm for space exploration, ultimately encourage NASA to build new infrastructure, and persuade Congress to fund similar missions. Guided by this philosophy, Perfect Space Development is working on a number of different space exploration projects, whose success will also boost Americans' awareness of personal space travel and pave the way for missions to Mars. The company is getting closer to its goal. In 2014, it completed the main part of the life support system in the Inspiration Mars mission, and conducted a complete test in a laboratory. From urine recycling and oxygen production to carbon dioxide removal, the system can provide all the conditions necessary for the survival of astronauts on the Inspiration Mars mission.

On October 26, 2015, WorldView successfully conducted a flight test using high-altitude balloons to enter space in Page, Arizona. During the test flight, the balloon carried a scaled-down simulated sightseeing capsule to a height of 30,624 meters. Currently, WorldView is still working on developing its Voyager capsule and booking long-haul flights. However, the company has not provided a specific timetable for high-altitude balloon travel, partly because it is focusing on its remote-controlled platform Stratollite – a large controllable high-altitude balloon that offers all the advantages of high altitude, low cost, rapid deployability, and low impact on flights. Compared with other high-altitude balloons, its innovation lies in the ability to fly in a variety of orbits, such as flying around the Earth or remaining aloft at a specific location for several weeks or months. The balloon platform can move slowly in a certain area for communication, remote-sensing, weather prediction, and scientific research tasks. Ultimately, the company also hopes to provide human space travel opportunities by charging a fee. On an exploratory mission, WorldView captured panchromatic images with a resolution of five meters at an altitude of nearly 23.5 kilometers above the stratosphere. This resolution is more than enough for tracking ground vehicles.

Poynter and MacCallum were children when the Apollo 11 landed on the Moon on July 20, 1969, but they remember it. It was this event that led them to Biosphere II and inspired their lifelong enthusiasm for space. They say: 'What we learned from Apollo was that a very difficult and inspiring technical program that involves humans exploring new worlds can create new heroes and role models, and inspire people into the sciences.'

5.4 The start-up mode in the aerospace industry

At present, the rising tide of global space entrepreneurship is spreading from the United States and Europe to China. Therefore, examples of various foreign space entrepreneurship models can provide a reference for Chinese space entrepreneurs.

The first model is elite cross-border entrepreneurship. The most successful of these are Elon Musk, founder of SpaceX, and Jeff Bezos, founder of Blue Origin. Both accumulated wealth during the initial Internet boom and, because of their aspirations for space, moved on to establish space companies. As well as bringing capital to their companies, they also inject entrepreneurial spirit, perspective, and ways of thinking that have been cultivated in the Internet field. Another example is Robert Bigelow, a real estate entrepreneur, who founded Bigelow Aerospace – a successful company that developed commercial space stations and hotels.

Members of the Chinese Internet elite are also concerned about the entry of private enterprises into the aerospace field. At the 2014 Chinese People's Political Consultative Conference (CPPCC) national committee meeting, Robin Li (the founder of Baidu) proposed to 'encourage private enterprises to enter aerospace industry field such as rockets and satellite launches to enhance the international competitiveness of China's space industry'. Tencent has also invested in commercial aerospace companies such as WorldView in the US. At present, although they are still observing and attempting to develop China's commercial aerospace industry, these companies will eventually devote themselves to the promising domestic commercial aerospace industry.

The second mode of entrepreneurship moves from campuses and research institutes to the market. One example is Skybox, which came out of Stanford University. The company started with CubeSats in 1999 (a 10 cm cube satellite), and soon turned to developing medium-sized satellites weighing 100 kg, achieving the ideal compromise between performance and cost. In addition, several of the founders of Planet Labs, which recently acquired Skybox, came from NASA's Ames Research Center. With their accumulated skills and networks in research centers, these individuals quickly succeeded in starting their own businesses. It seems that this kind of space entrepreneurship mode is the most suitable for China, and most of the space entrepreneurship teams in China belong to it.

The third entrepreneurial model involves amateurs and enthusiasts going to market. Poynter and MacCallum are examples. From young amateurs to successful entrepreneurs, they have weathered a difficult path. Another team, Copenhagen Suborbitals, led by several young Danish scientists, is a non-profit organization funded entirely by donations. Their ideal was to build their own rocket that would be capable of carrying a person into sub-orbit, and many rocket and spacecraft tests have been carried out.

There are many aerospace amateurs and enthusiasts in China, and some have entered the ranks of aerospace entrepreneurs. If they want to succeed, they should also be prepared for a long struggle.

At present, the strength of China's space entrepreneurs is very different from that of state-owned aerospace enterprises. It is not on the same scale at all. However, a journey of a thousand miles begins with a single step, and those who dare to try will provide valuable lessons whether they succeed or not. With the consistent influx of private capital, many experienced scientists and managers in the system are sure to join the tide of space entrepreneurship.

References

[1] Huang Zhicheng. *Sky and Sky Vision* [M]. Beijing: Electronic Industry Press, 2015.

[2] Ashlee Vance. *Elon Musk: How the Billionaire CEO of SpaceX and Tesla is Shaping our Future* [M]. Zhou Hengxing, trans. Beijing: CITIC Publishing House, 2016.

[3] Huang Zhicheng. 'Why Chinese Aerospace Needs Private Enterprises' [J]. *International Space*, 2014 (11): 14–15.

[4] Huang Zhicheng. 'Security: the lifeline of private space travel' [J]. *International Space*, 2014 (12): 14–15.

[5] Huang Zhicheng. 'The pioneering wave of aerospace development' [J]. *Space Exploration*, 2015 (1): 20–26.

[6] Huang Zhicheng. 'The Great Power Game Behind the Growth of SpaceX' [J]. *Military Digest*, 2016 (3): 6–8.

[7] Huang Zhicheng. 'Behind the success of the 'American Madman'' [N]. *Wenhui Bao*, 2016–01–10 (7).

[8] An Hui. 'The 'Dragon' spacecraft from nursery rhymes' [J]. *Space Exploration*, 2012 (7): 46–49.

[9] Huang Zhicheng. 'Musk's success is not down to luck' [N]. *China Youth Daily*, 2014-05-12 (2).

[10] Huang Zhicheng. 'A Martian couple' [J]. *Space Exploration*, 2015: (2) 30–33.

[11] Huang Zhicheng. 'The United States begins the commercial era of manned spaceflights' [J]. *Space Exploration*, 2014 (11): 17–21.

[12] Zhang Rui. 'The Commercialization of American Manned Spaceflights' [J]. *Spacecraft Engineering*, 2011, 20 (6): 86–93.

[13] Zhang Rui. 'The Commercialization of American Manned Spaceflights (Part 1)' [J]. *International Space*, 2016 (4): 62–68.

[14] Zhang Rui. 'The Commercialization of American Manned Spaceflights (Part 2)' [J]. *International Space*, 2016 (6): 44–48.

[15] Huang Zhicheng. 'Spacecraft 1 begins a new chapter in the history of space travel' [J]. *Space Exploration*, 2004 (8): 22–25.

[16] Tao Tao. 'Skybox's high-resolution small and micro satellite constellation begins the era of commercial remote-sensing 2.0' [J]. *Satellite Applications*, 2014 (3): 70–71.

[17] Federal Aviation Administration. The Annual Compendium of Commercial Space Transportation: 2017 [R/OL]. https://brycetech.com/downloads/FAA_Annual_Compendium_2017.pdf.

CHAPTER 5

THE FUTURE SUCCESS OF COMMERCIAL REMOTE-SENSING SATELLITES

New business concepts, financing tools, and management modes have reduced the entry threshold of commercial remote-sensing satellites and promoted its rapid industrialization. At the same time, the commercial market demand has expedited the progress of remote-sensing technology, and accelerated technological innovation and integration. Looking to the future, the combination of remote-sensing satellite technology and other cutting-edge technologies will promote the further development of commercial remote-sensing satellites and become a new aerospace industry.

1. The History of Commercial Remote-sensing Satellites

Since the successful commercial operation of IKONOS-2 satellite in 1999, many commercial remote-sensing satellites have been launched around the world. These satellites have dual military and civilian uses, and are operated in the market in both global and local areas. As well as opening huge market space, commercial remote-sensing satellites also greatly promote the rapid progress of the entire remote-sensing satellite system, and encourage the healthy and steady development of the whole industrial chain.

1.1 The rise of commercial remote-sensing satellites

In 1972, Landsat-1 – the world's first remote-sensing satellite – was successfully launched, beginning the development of the industry. With the maturity of technology and the expansion of application, military and government departments have increasing demands for high-resolution remote-sensing satellite images. Western aerospace powers represented by the United States and France have rapidly developed and commercialized high-performance remote-sensing satellites through pro-active policy guidance and financial support, forming a virtuous cycle of business models for government supervision and independent operation by enterprises. Most commercial remote-sensing satellites serve the government and defense users. Meanwhile, they continue to develop and expand in the civilian market, becoming an example of the integration of military and civilian development for aerospace.

After the Cold War, the United States sold declassified reconnaissance satellite images for commercial purposes. In 1992, US Congress passed the Land Remote-sensing Policy Act, which clarified the importance of commercial remote-sensing images to national security, and authorized the US Department of Commerce to issue licenses for private enterprises to operate remote-sensing space systems. In 1994, then president Bill Clinton issued the PDD-23 US Policy on Foreign Access to Remote-sensing Space Capabilities, which allowed private companies to take high-resolution images from space, and began a new era of commercial remote-sensing satellites. On April 25, 2003, then president George W. Bush approved a new policy on commercial remote-sensing, promising that the government would use commercial systems to meet its need for imagery and geospatial information, encourage the industry to build more optimized commercial remote-sensing systems, and establish interactive mechanisms between commercial remote-sensing and national interests.

IKONOS-2 was the first commercial remote-sensing satellite, while the QuickBird was the world's first commercial satellite to provide sub-meter resolution data, and the WorldView4 is currently the world's highest-resolution commercial remote-sensing satellite. All of these satellites are milestones in the development of commercial remote-sensing. At the same time, service systems for commercial remote-sensing satellites were established with representatives from the WorldView series in the United States, the SPOT series in France, and the COSMO-SkyMed series in Italy.

1.2   The first generation of commercial remote-sensing satellites in the USA

Resolution is the key technical index of remote-sensing satellites, and generally refers to the geometric resolution ability of a remote-sensing device to ground objects. It can also be understood as the minimum distance between two targets on the ground in remote-sensing images. High-resolution images from remote-sensing satellites have long been restricted by military authorities in countries such as the USA and the Soviet Union/Russia, blockading commercial access.

The first high-resolution commercial remote-sensing satellite was called IKONOS-2 and it was developed from the Space Imaging Inc. joint venture that included Lockheed

Martin, Raytheon, and companies from Europe, Japan, South Korea, and Singapore. Taking its name from the Greek word meaning 'images', IKONOS was built by Lockheed Martin. It was launched into orbit on September 24, 1999 from Vandenberg AFB aboard an Athena II launch vehicle. The satellite is currently in polar orbit at an altitude of 680 km and an inclination angle of 98.2°. It was designed to circle the Earth 2,049 times in 140 days (about 15 times per day), and the sub-satellite point of first circle is exactly as same as the 2,049[th] circle. Every three days, any area on the ground can be scanned at a resolution of 0.8 meters. If the resolution is reduced, same area can be revisited once a day. The coordinates of the targets are uplinked to a computer sorting system, which transmits the tasks into the satellite from ground stations. A ground station can immediately receive the downlinked imagery. The satellite's camera weighs 170 kilograms and has a power of 350W, and was manufactured by KODAK, which has long provided cutting-edge optical equipment to the National Reconnaissance Office (NRO). The Optical Telescope Unit (OUT) on the satellite captures images along an 11-kilometer-wide strip, and reflects them onto a digital imaging remote sensor. Panchromatic (PAN) and multispectral (MS) imaging sensor arrays mounted in an FPU (Focal Plane Unit) convert light into imaging cells. The FPU is also equipped with A/D converters that transform each image cell into data bits. A DPU (Digital Processing Unit) then compresses and formats the digitized image and transmits it back to the ground. The IKONOS-2 satellite can capture panchromatic (black and white) images of objects less than one meter in diameter on the ground, and multispectral images of objects with a diameter of only 3.28 meters. That is an improvement on France's early Helios reconnaissance satellites, which had a panchromatic resolution of just two meters.

The IKONOS-2 satellite has ground stations in Alaska, Oklahoma, Sweden, Greece, and Thornton (Colorado), where the control center and headquarters of Space Imaging are also located. The processing and delivery of satellite images are very convenient. In digital forms, Space Imaging's digital image data can be sold to users in the form of raw data, or can be sold together with software or other data to form enhanced products. On March 31, 2015, the IKONOS satellite was decommissioned after 15 years, and its working time was more than twice its design life.

In December 2000, the National Oceans and Atmospheric Administration (NOAA) approved Space Imaging and EarthWatch to operate imaging satellites at a resolution of 0.5 meters. EarthWatch launched the EarlyBird-1 – a high-resolution remote-sensing satellite with a panchromatic resolution of three meters – in December 1997, but it was scrapped after just four days in orbit. This led to layoffs. In November 2000, the company launched the QuickBird I satellite from the Plesetsk Cosmodrome using the Russian Kosmos carrier rocket, which also ended in failure. In September 2001, EarthWatch changed its name to DigitalGlobe, and on October 18, 2001, it successfully launched the QuickBird II satellite with Boeing's Delta II rocket. The satellite is located in a sun-synchronous orbit with a height of 450 km and an inclination of 98°. As the satellite reduces the orbital height of the original design, it provides industry-leading remote-sensing data at 0.61 m panchromatic (black and white) and 2.44 m multispectral (visible, infrared). The main advantage of the QuickBird II is that it can take images at high resolution of one meter with a width of 22 kilometers, while the Space Imaging has a satellite imaging width of 11 km. Several companies participated in this program, including Ball Aerospace, Hitachi, European Imaging, and International Telephone & Telegraph (ITT).

The Orbital Imaging Corporation (ORBIMAGE) is a subsidiary of Orbital Sciences Corporation (OSC). Its OrbView-3 high-resolution imaging satellite, equipped OHRIS (OrbView High Resolution Imaging System) with 1m-resolution panchromatic images and 4m-resolution multispectral images, was successfully launched by a PEGASUS carrier rocket from OSC. Together with two lower-resolution OrbView satellites launched in 1995 and 1997, they will conduct a variety of weather-related observations for NASA and other users in the United Status.

1.3   The Military Value of Commercial Remote-sensing Satellites

Although high-resolution remote-sensing satellite technology has long been controlled by the US military, valuable military information can also be obtained from high-resolution commercial satellite images, while commercial applications have to be coordinated with relevant departments of the US government. In fact, from these images, missiles and artillery can be found, troops can be identified, airports and missile bases can be located, troop concentrations and dynamics can be monitored, large-scale military maps can be made, and routes for military operations can be selected. A large aspect of the initial business of commercial remote-sensing satellite companies came from the US intelligence community. Of course, some countries that do not have space reconnaissance capabilities have to buy high-resolution images to maintain their national security. High-resolution commercial satellites are obviously less capable than the reconnaissance satellites of the National Reconnaissance Office (NRO). The resolution of NRO satellites can reach 0.3 meters, but their images are strictly confidential. The Clinton administration issued a Presidential Decision Directive on the commercial remote-sensing system (NSC/PDD-23), stipulating that 'During period when national security or international obligations and/or foreign policies may be compromised, as defined by the Secretary of Defense or the Secretary of State, the Secretary of Commerce may, after consultation with the appropriate agency(ies), require the licensee to limit data collection and/or distribution by the system to the extent necessitated by the given situation. Decisions to impose such limits will only be made by the Secretary of Commerce in consultation with the Secretaries of Defense or the Secretary of State, as appropriate. Disagreements between Cabinet Secretaries may be appealed to the President.'

In fact, after the September 11 attacks, the United States adopted the purchasing of exclusive rights to control the imagery of high-resolution commercial satellites. On October 11, 2001, the National Imagery and Mapping Agency (NIMA) (now called the National Geospatial-Intelligence Agency) of the United States signed an agreement with the Space Imaging Corporation to purchase the exclusive right to all images taken by the IKONOS-2 satellite over the war zone in Afghanistan, and paying the company 1.9 million dollars per month.

2. The New Generation of Commercial Remote-sensing Satellites

In recent years, the growing popularity of the Internet has provided an ideal path for the distribution of digital image products. With the consistent development of remote-sensing satellite technology, a new generation of market-oriented commercial remote-sensing satellites have emerged.

2.1 The NextView Program

Sub-meter resolution optical imaging satellites such as the WorldView series and the GeoEye series are the most powerful in-orbit commercial optical imaging satellites in the world. The development of these satellites has a military background, and more than half of their business applications still serve the US military.

The National Geospatial-Intelligence Agency (NGA) is the driving force behind the American high-resolution commercial optical imaging satellites that are currently in orbit. The NGA, formerly known as the National Imagery and Mapping Agency (NIMA), was formed in 1996 by the National Photographic Interpretation Center (NPIC) of the Central Intelligence Agency (CIA), the office of Defense Intelligence Agency (DIA), the Image Analysis Office of the DIA, the Central Imagery Office (CIO), and the Defense Dissemination Program Office. It is part of the US Department of Defense (DOD). Its primary mission is to provide strategic and tactical image intelligence to military decision-makers and military operations in a timely and accurate manner, while meeting the needs of national decision-makers and other non-military sectors. In addition, it performs certain functions of agencies such as the DIA, the National Reconnaissance Office (NRO), and the Defense Airborne Reconnaissance Office (DARO). With the signing of the National Defense Authorization Act for Fiscal Year 2004, the NIMA officially changed its name to the NGA. The NGA is a key component of the US intelligence community and an important operational support organization. It accepts the dual leadership of the CIA and the DOD. Its main task is to research imagery data from US military reconnaissance satellites and high-resolution commercial remote-sensing satellites. It also identifies ways to map out the corresponding products and provide geospatial information including analysis of the Earth's natural characteristics and geographic benchmarks, so as to support national security.

In October 2003, the NGA, using the name 'NextView program', funded DigitalGlobe to develop the WorldView series of satellites in the form of pre-paid imagery purchasing costs to replace the 0.6-meter resolution QuickBird II satellite. The WorldView-1 and WorldView-2 satellites were successfully launched into orbit in 2007 and 2009 respectively. In September 2004, as part of the NextView program, NGA funded what was then known as the Orbital Imaging Corporation (ORBIMAGE) to develop the OrbView5 satellite in the same way. In 2006, ORBIMAGE was reorganized as GeoEye, and the OrbView5 satellite was renamed GeoEye-1. This satellite was successfully launched in 2008 with an image resolution of 0.41 meters.

The NGA signed a 10-year contract for EnhancedView remote-sensing satellite data procurement with DigitalGlobe and GeoEye in 2010, worth 3.5 billion and 3.8 billion US dollars respectively, to support the development of GeoEye-2 and WorldView-3 satellites with higher resolution. Due to the tightening of the US space budget in 2012, the NGA reduced the contract amount, leading to the merger of GeoEye by DigitalGlobe, in which the GeoEye-2 satellite was renamed WorldView-4.

In 2013, DigitalGlobe formally applied to the US government to sell satellite image data with a resolution of 0.25 meters. In June 2014, the US Department of Commerce (which was responsible for issuing commercial remote-sensing satellite operation licenses) officially approved DigitalGlobe's application with the approval of the Defense of Department and the State Department.

2.2 The WorldView-3 Satellite

In addition to the pursuit of high-precision and high-resolution imaging performance, the WorldView satellite series also requires a significant increase in the satellite's rapid attitude maneuverability to increase the range of observations and shorten the revisit cycle (the WorldView-1's revisit period is 1.6 days) as well as providing rich working modes to improve the application performance of the satellites. Based on the BCP-5000 satellite platform of Ball Aerospace & Technologies Corp., the WorldView satellites are equipped with advanced Control Moment Gyroscopes (CMGs), Star trackers, precision IRU (Inertial Reference Units) and GPS receivers, to precisely control the attitude. This gives the satellites strong Off-Nadir ability, and can provide sub-satellite point imaging, multiple target imaging in the same orbit, stitched imaging, multi-view stereo-imaging of the same orbit, and other working modes. The resolution of the spaceborne camera is greatly improved by an optical system that features a large aperture and long focal length, and a panchromatic multispectral imaging device is adopted to obtain richer and more reliable image information.

DigitalGlobe launched the WorldView-3 satellite in August 2014. Six months later, the company began selling panchromatic images worldwide with a resolution of 0.31 meters. WorldView-3 is the first commercial satellite to feature super spectral and super resolution. Generally speaking, when the band of a satellite exceeds 10, it is a hyperspectral remote-sensing satellite. The WorldView-1 has only one panchromatic band, while the WorldView-2 has one panchromatic band and eight multispectral bands. The WorldView-3 has one panchromatic band, eight multispectral bands, eight short-wave infrared bands (SWIR), and 12 CAVIS bands.

In addition to enhanced spectra, the WorldView-3 satellite also carries brand new sensors to detect clouds, ice, snow, fog, and aerosols, and uses 12 atmospheric correction bands (CAVIS). For typical maps or atmospheric disturbance factors, it has corresponding bands according to its characteristics. The CAVIS (Clouds, Aerosols, Vapors, Ice, and Snow) band will receive ground information synchronously with the high-resolution detector, which can not only obtain high-resolution and hyperspectral resolution ground image data, but also obtain information on different characteristics. This can greatly improve the accuracy of spectral image analysis, and even achieve quantitative weather remote-sensing through the CAVIS band.

The spatial resolution of the panchromatic band of the WorldView-3 satellite is 0.31 meters, that of the multispectral band is 1.24 meters. he resolution of the short-wave infrared band is 3.7 meters, and that of the CAVIS band can reach 30 meters.

In terms of acquisition capability, the WorldView-3 satellite has increased its acquisition unit by 20%. With the same resolution, the acquisition capability of the WorldView-3 is increased by 20%, and the downlink data speed is increased by 50%.

2.3 The WorldView-4 Satellite

DigitalGlobe successfully launched the WorldView-4 satellite in November 2016. Compared to the WorldView-3, it can move from one target to another faster, and can store more data. The successful launch of the WorldView-4 satellite once again greatly improved the overall data acquisition capability of DigitalGlobe's constellation, enabling it to take photos of any location on the Earth at an average frequency of 4.5 times per day.

Design & Specification of the WorldView-4

- Orbit: Altitude: 617 km

  Type: Sun Synchronous orbit

  Period: 97 min.
- Estimated service life: 10 to 12 years
- Spacecraft size and aperture: Size: 5.3 m (17.7 ft) tall × 2.5 m (8 ft) across 7.9 m (26 ft) across deployed solar arrays; Aperture: 1.1 m
- Sensor resolution: Panchromatic Nadir: 0.31 m, Multispectral Nadir: 1.24 m
- Swath width: At nadir: 13.2 km
- Ultra-large storage capacity under various image acquisition modes
- Capable of performing rapid re-orientation maneuvers from one imaging target to the next, at more than twice the speed of other satellites by Control Moment Gyroscopes (CMGs), providing more optimized image acquisition capability for regional and point targets
- Delivers a variety of data products directly to users
- Average revisit time of < 1 day

In addition to providing real-time hyperspectral and high-resolution Earth images, the WorldView-4 satellite also has the following incomparable advantages: it can conduct large area single-pass (synoptic) collection, eliminating temporary variations, and its global capacity is 680,000 square kilometers per day. With the launch of the WorldView-4 satellite, DigitalGlobe has made improvements to its systems, platforms, software, and team that will enable end users to solve more problems with greater accuracy and deliberation.

The USA's WorldView and QuickBird satellite series are high-resolution constellations. Since the launch of the first QuickBird satellite in 2001, six satellites have been launched and put into use. Launched on September 6, 2008, the WorldView-4 satellite and the GeoEye-1 satellite are the same kind, and form constellations. The successful launch of the WorldView-4 satellite has given DigitalGlobe a near monopoly on the high-resolution commercial image market in the United States. It has become a global supplier of commercial high-resolution Earth imagery products and services, providing images and solutions that help decision-makers to understand the changing planet in order to save lives, resources, and time. With data from satellite constellations, DigitaGlobe's image solutions can provide customers with coverage information that is tailored to their needs.

As of June 2016, DigitalGlobe's revenue was 175.4 million US dollars, of which the US government procurement contracts accounted for 84.3 million. Currently, DigitalGlobe has 11 Direct Access Partner (DAP) users worldwide who can control satellites and direct data distribution locally. DigitalGlobe restricts the distribution areas of DAP users according to their contract. In the process of building satellite constellations and developing new algorithms, DigitalGlobe is consulting customers for new experiences that are different from traditional data services.

2.4 Examples of Other Commercial Remote-sensing Satellites

In 2016, DigitalGlobe and Saudi Arabia's Taqnia Space established a joint venture company with King Abdulaziz City for Science and Technology (KACST) to build at least six small optical imaging satellites. The company offers products, sales, and marketing. KACST will be responsible for manufacturing, integrating, and launching satellites, and will own 50 percent of the imaging capacity for Saudi Arabia and the surrounding area. DigitalGlobe will own the other 50 percent of the imaging capacity, as well as full imaging capacity for the rest of the world.

At present, four commercial optical imaging satellites in orbit have been networked and operated by Airbus Defense and Space. The constellation has the ability to revisit twice a day, and the SPOT satellite provides a wide image with a resolution of 1.5 meters and a swath width of 60 kilometers. Launched in 2011 and 2012, the Pleiades series offers precise images with a resolution of 0.5 m and a swath width of 20 km for specific target areas. The Pleiades satellites have established military ground stations in France and Spain, and civilian ground stations in France and Sweden. Other major users have deployed mobile and fixed ground stations. At present, SPOT7 has been acquired by Azerbaijan as a real user. Prior to this, SPOT provided a series of training schemes, including the establishment of personnel and ground facilities.

Europe ImageSat International (hereinafter referred to as ISI) is an end-to-end geospatial information solution provider that delivers services including satellite constellations, ground station networks, data services, mapping and GIS services, battlefield situational awareness, agriculture & forestry, and intelligence analysis. By 2012 it had launched EROS-A and EROS-B (with a resolution of 0.7 meters) to form a satellite constellation to improve revisiting capabilities.

UrtheCast, a Canadian geospatial information provider, announced that it will add a constellation of eight satellites to the 16 satellite systems it launched in 2015, and is now promoting high-resolution satellite imagery to users worldwide through its website.

3. New Commercial Remote-sensing Satellite Companies

When the aerospace industry was in its infancy, the high cost of building and launching satellites discouraged many institutions and companies. The Solar Dynamics Observatory satellite launched by NASA in 2010 weighed three tons, and the cost of preparation and launch was about 850 million US dollars. With the progress of satellite miniaturization technology, this situation has begun to change. For example, in 2010, NASA and the Department of Defense launched the Fast, Economically Affordable Science and Technology Satellite (FASTSAT), which weighed only about 182 kilograms and cost to about 10 million US dollars. With the development of commercial aerospace, even greater changes will take place in the satellite field.

3.1 Miniature & personal satellites

Peter Klupar, Director of Engineering at NASA's Ames Research Center, used to pull out his Blackberry at meetings and ask: 'Why do smartphones have faster processing capabilities and better sensors than many satellites? And why are their costs relatively low?' before putting the phone back in his pocket and continuing the meeting.

Will Marshall and Chris Boshuizen, researchers at the Ames Research Center, often talked about Klupar's confusion. They started discussing the cost of satellites, and began to think about the possibility of embedding phones into them.

Marshall, Boshuizen, and their colleague Robbie Schingler established the Small Spacecraft Office at Ames. They helped to establish a nonprofit organization called the Space Generation Advisory Council (SGAC), made up of a group of young professionals hoping to inject new thinking into aerospace research. 'Consumer electronics and other parts of the tech industry have developed a lot of technologies that are useful for satellites, but many people in the sector have largely ignored that, because they have their own set of things,' Marshall said.

PhoneSat was developed at the Ames Research Center with the support of NASA's Small Spacecraft Technology Program. PhoneSat is a CubeSats unit with standard dimensions of 10 cm × 10 cm × 10 cm. It typically weight less than 1.33 kilograms per unit, and runs Google's Android operating system. It comes in two versions, namely PhoneSat 1.0 and PhoneSat 2.0. The former was modified by HTC's Nexus One smartphone with a radio transmitter, battery, and watchdog circuit. The watchdog circuit is used for simple monitoring of satellite systems and for restarting the smartphone when the radio stops sending data packets. PhoneSat 2.0 is modified with a better Nexus S smartphone made by Samsung Electronics, including a two-way S-band radio antenna, triangular solar panels, GPS receivers, and attitude control mechanisms such as magnetorquer coils and reaction wheels.

To protect the smartphone, the PhoneSat 1.0 satellite placed the phone inside a 10 cm cube with a weight of about 1.8 kg. In addition to smartphones, all other components of this 'phone satellite' are commercial off-the-shelf (COTS) without any modification. As a result, the cost of each 'phone satellite' is only about 3,500 US dollars.

In April 2013, the first three PhoneSat satellites (two version 1.0 models and one version 2.0) were launched into circular orbit with an inclination of 51.6 degrees and an altitude of 250 km by the Orbital Sciences Corporation's Antares rocket. The three phone satellites stayed in orbit for nearly a week to carry out tests such as collecting telemetry data, taking pictures of Earth with the smartphone camera, and conducting data communication with the Iridium constellation. On November 19, 2013, PhoneSat 2.4 (weighing about 1 kg and with a lifespan of about two years) was launched into orbit with communication transmission capabilities, and an ability to control the attitude of the satellite through the reaction wheel. It is the first PhoneSat satellite to use S-band radio. The PhoneSat 2.5 was launched on April 18, 2014 to test radio communications and positioning systems.

According to NASA, smartphones can provide many of the capabilities needed for satellite systems, including high-speed processors, powerful operating systems, a variety of tiny sensors, high-resolution cameras, GPS receivers, and a range of radio communications equipment. Phone satellites are now ready to be used for future space projects such as lunar exploration, low-cost Earth observations, and testing new technologies and components for space flight.

Two decades ago, satellite technology sparked a miniaturization revolution. Faced with a shrinking budget today, NASA is having to reduce expenditure in many aspects, and the phone satellites have emerged as the times require. The purpose of the program is described in the PhoneSat Flight Demonstrations as 'launching the lowest-cost and easiest to build satellites ever flown in space.' Obviously, the significance of this program is not only a question of technology, but more importantly of reducing the cost of space exploration by space institutes in the United States. This means that satellite technology will enter ordinary households, beginning an era of personal satellite ownership.

3.2 Smart and Agile Planet Labs

In 2011, Marshall and two others who had worked on the PhoneSat program left NASA to start Planet Labs – a company focused on small and low-cost satellites. In explaining their idea for the satellite construction process, they stated that the company would not follow the approach of the Apollo Moon landing program, through which many systems were designed and analyzed before these systems were built. They had decided to adopt a software design method instead, which would first build a minimal viable prototype, and then proceed to satellite manufacturing. They called this idea 'Smart Space', and pointed out that most satellite parts could actually be found in mobile phones. If you made a component list of a satellite and a smartphone, you would find many of the same parts: GPS, radio, hard drive, CPU, and accelerometer.

Planet Labs satellites are called 'Dove', and their plan is either to release a large group of satellites in orbit or build a constellation called Flock-1. The constellation consists of more than 200 CubeSats, each made up of three units. The size of each satellite is 10 cm × 10 cm × 30 cm, with red, green, blue, and infrared bands, and an imaging resolution of 3–5 meters. The ground station adopts the UHF band for measurement and control, and the X-band for data transmission to ensure the stable operation and data reception of the satellite. In February 2014, they released Flock-1 CubeSats (including 28 Earth-observing satellites) from the ISS, and then deployed three more satellites. On February 15, 2017, India successfully launched 104 satellites in a single launch, including 88 Flock-1 satellites. The number of Flock-1 satellites in orbit reached 119, and they take full-coverage pictures of the Earth every day. On July 18, 2017, Planet Labs launched another 48 Flock-2 satellites with a Russian Soyuz rocket.

Planet Labs offers open and free access to its imaging data, and has applied its technology to the industry sectors that need remote-sensing satellite data, such as agriculture, transport, and mining. These satellites can show deforestation and melting ice sheets in real time, providing accurate data for climate change research. They can also predict, witness, and warn of natural disasters such as floods, fires, hurricanes, and tsunamis, helping with disaster relief. An earthquake measuring 7.8 on the Richter scale struck Nepal in April 2015. After digging into its own data, Planet Labs found that rescuers had overlooked two remote villages. It immediately alerted rescuers, who were able to bring life-saving supplies to these villages.

Planet Labs is typical of an organization in the new space age. It is a new-generation Silicon Valley start-up for commercial aerospace based on the Internet, and is growing rapidly alongside Internet enterprise development. Thanks to the consistent injection of capital from Silicon Valley investors and its own strength, it has put more than 100 satellites into orbit, and has merged with some of its biggest rivals in the industry. In July 2015, Planet Labs acquired BlackBridge, whose RapidEye constellation gives the company the ability to quickly access a wide range of multispectral data. On April 18, 2017, Planet Labs acquired Terra Bella (formerly known as Skybox) from Google, and acquired all of its business, as well as seven in-orbit SkySat satellites. Since then, Planet Labs has acquired high-resolution image collection capabilities as well as Terra Bella's core remote-sensing data mining team. No less, Google has signed a long-term imagery data contract with Planet Labs. According to sources, it is an equity deal that could result in Alphabet also owning a stake in Planet Labs.

A core factor in Planet Labs' success is that it uses Internet thinking to change the satellite industry. It has worked at speed, not least in the rapid prototyping of products (it developed its first satellite in less than a year) and the fast iteration of a technology known as Agile Aerospace. In the few years since the first satellite was developed, the system has been upgraded more than 10 times. Early satellites had lower image resolution, few imaging bands, and no attitude control. Nowadays, the ability of satellites launched into orbit has been greatly improved. They have better sensors with higher imaging capabilities, the same batteries as those in Tesla cars, and more complete controls, from magnetic torques to reaction wheels. Although the weight and size of the satellite has not changed significantly, rapid technological iteration has greatly improved its capability. Such speed of iteration is not surprising for Internet companies, but it is simply unimaginable in the traditional satellite industry.

A few years ago, the industry agreed that CubeSats should only be used for technical verification and teaching experiments due to their low practical value. The success of Planet Labs has overturned this notion. With the rapid development of consumer electronics, the latest cutting-edge electronic components have been used in CubeSats satellites. With its five to six kilogram CubeSats satellites, Planet Labs has successfully occupied the market for middle and low resolution remote-sensing imagery.

Planet Labs' development concept has redefined the risk awareness of commercial aerospace. In the past, after repeated tests and trials, satellites were able to go into orbit. Today, however, Planet Labs advocates that satellites should be built as soon as possible and then be launched into orbit for verification. Planet Labs has indeed experienced many failures, but it has developed quickly due to the lower cost and rapid iteration of CubeSats, the same rapid iteration of upstream components, and its flexible, low-cost, and efficient launch. This new risk control approach integrates into the satellite development process, and is also being accepted by investors whose confidence will not be affected by failure. There is no doubt that Planet Labs is now beginning a new era of rapid development for commercial aerospace.

3.3 Orbital Insight's application innovation

Data mining and in-depth application and analysis of mass remote-sensing have become important factors in promoting the development of commercial remote-sensing satellites. As a result, a number of companies engaged in remote-sensing satellite data service and innovative application have emerged, including Orbital Insight. Its founder, James Crawford, has worked on Artificial Intelligence (AI) systems at NASA and other agencies, and was once the engineering director at Google Books. So far, the company has raised a total of 78.7 million US dollars.

Relying on remote-sensing images from the world's leading satellite image providers, Orbital Insight uses Deep Learning technology to identify objects and features in images, to display data results intelligently, and to convert pixels into measurable data for in-depth analysis by data mining technology. Their algorithms can calculate and measure roads, airplanes, clouds, smoke, land, buildings, cars, and oil tanks to gain a better understanding of the world. The company is now providing remote-sensing image analysis services to more than 60 capital asset management firms, including some government sectors, institutions, and two global non-profit organizations, in the analysis of crop satellite image data, prediction of grain production, parking lot vehicle data, and retail sales prediction.

When Orbital Insight learned how to count cars through its Deep Learning program, the company was able to determine the digital relationship between the number of cars in supermarket parking lots and the retailer's revenue in a given quarter. Furthermore, more specialized indicators can be obtained by analyzing trends in a certain country or the world. For example, by observing the shadow changes of buildings in a region of China through satellite images, the development status of China's construction industry can be measured to find out whether the speed of construction is increasing or not. By looking at satellite data from oil storage tanks around the world, users can immediately identify how oil supply has changed over a given period of time, while current methods typically have a lag of several weeks. Orbital Insight has also worked with the World Bank to improve the accuracy of its data on social poverty by analyzing building heights and roofing materials to determine the wealth of residents.

Orbital Insight's founder James Crawford has said, 'In the long term, I see this as a way to understand the world at scale. The combination of satellites and software allows people to really see and assess the world around them.'

3.4 New Trends for Commercial Remote-sensing Satellites

With the commercialization of high resolution remote-sensing satellites, Earth image data is widely used in fields such as agriculture, environmental monitoring, mapping, urban and regional planning, oil and gas development, communications, disaster prevention, automobile navigation, and real estate sales. For example, farmers can use satellite images to monitor crops, while telecommunications companies and power sectors can use images to locate the towers of base stations or select transmission lines. Real estate agents can use them to give customers a preview of future properties and their surroundings.

Global sales of commercial remote-sensing data totaled 2.3 billion US dollars in 2015, resulting in an output value of nearly 6 billion US dollars for the global geographic information industry, of which 83% was optical data. Looking ahead to the commercial remote-sensing data sales market, Europe and the United States have the major share, but demand has been growing rapidly in Asia, Latin America, Africa, and the Middle East. DigitalGlobe (DG) and Airbus Defence and Space (ADS) account for 79% of global sales. In the process of rapidly promoting the industrialization of remote-sensing satellites, some new trends have emerged.

Firstly, the development of commercial remote-sensing satellites and small satellite technologies is driving major changes in the satellite field. Some scholars refer to this change as the 'Orbital Revolution'. The core content of this revolution is that the rapid development of micro satellites and low-cost rockets has greatly reduced the threshold of entering and utilizing space, and also pushes space technologies and activities into a new era of popularity. For commercial remote-sensing satellites, new technologies such as microelectromechanical systems, 3D printing, new materials, and new energy applications have greatly reduced the weight of satellites, extended their life span, and reduced costs. The integrated application of the high-precision Control Moment Gyroscope (CMG), star tracker, precision IRU (Inertial Reference unit), and GPS receivers greatly improves the positioning accuracy of satellites and the resolution of Earth observation. It is precisely because of the reduction of cost that some satellites can use ordinary electronic components to reduce costs and shorten the manufacturing time. Looking to the future, commercial remote-sensing satellites need to improve their on-board data processing capacity, enhance the data transmission speed between satellites and ground stations, and eradicated the heat dissipation of electronic components.

Secondly, cutting-edge technologies are constantly expanding new applications in the field of remote-sensing satellites. In specific applications, high-resolution images from commercial remote-sensing satellites should be combined with aerial photographs, the Global Positioning System (GPS), the Geographic Information System (GIS), computer processing, and the Internet. They should be supported by advanced technologies such as Cloud computing, Big Data, AI, and VR, which enable users to analyze data and solve many real social problems. For example, satellite images can extract residential and road networks accurately to optimize navigation functions, thereby improving route planning instructions and enhancing the overall user experience. They can also help to build 5G networks to support the development of smart cities and networked vehicles; monitor and assess disasters or invasive species; track and record illegal fishing and poaching; detect and monitor deforestation or forest resources; calculate and measure changes in global economic indicators; and monitor the illegal use of pipelines and public utilities.

Finally, capital plays an increasingly important role in the development of the commercial remote-sensing satellite industry. Planet Labs acquired Terra Bella from Google in 2017. In February 2017, McDonald Dettwiler and Associates (MDA) announced its acquisition of DigitalGlobe. In July 2017, MDA announced that its subsidiary, Laura Space Systems, had signed a contract with DigitalGlobe to build the next generation of the high-resolution remote-sensing satellite constellation WorldView series for DigitalGlobe. The constellation greatly enhances DigitalGlobe's ability to collect high-resolution images in important areas. Combined with high resolution and fast revisit capability, this constellation will join the existing WorldView series and the soon-to-be-deployed Scout small satellite constellation. DigitalGlobe will be able to collect imagery on the fastest-changing regions on Earth from sunrise to sunset every 20–30 minutes. The first satellite is scheduled to be launched in 2020.

New management concepts, financing means, and management modes have reduced the entry threshold of commercial remote-sensing satellites, increased the enthusiasm of private firms and capital to participate in commercial remote-sensing, and promoted the industrialization of commercial remote-sensing satellites. At the same time, commercial market demand has promoted the progress of remote-sensing technology, and has accelerated technological innovation and integration. Looking to the future, the combination of remote-sensing satellite technology with cutting-edge technologies such as Cloud computing, Big Data, artificial intelligence, and virtual reality will promote the further development of commercial remote-sensing satellites.

References

[1] Huang Zhicheng. *The Fourth Wave of Aerospace Science, Technology, and Society* [M]. Guangzhou: Guangdong Education Press, 2007.

[2] Zhuo Peng. 'Several new concepts in remote-sensing applications' [J]. *China Aerospace*, 2002 (1): 1–9.

[3] Huang Zhicheng. 'Digital Earth—A Common Digital Home for Mankind' [J]. *Qiushi*, 2000 (18): 60–61.

[4] Huang Zhicheng. 'Small and miniature modern satellites' [J]. *Science & Technology Review*, 1998 (2): 10–13.

[5] Huang Zhicheng. 'A new high-resolution commercial remote-sensing satellite' [J]. *Chinese Journal of Image and Graphics – B Edition*, 2001 (2): 15–20.

[6] Gong Ran. 'The US remote-sensing satellite policy and regulatory system, and its role' [J]. *Satellite Applications*, 2013 (3): 25–30.

[7] Zhang Shaohua, Xu Dalong. 'The Development of American Commercial Remote-sensing Satellites' [J]. *Surveying and Spatial Geographic Information*, 2016 (12): 135–138.

[8] Zhou Runsong. 'The Development of America's Super-High Resolution Commercial Optical Imaging Satellites and Data Service Models' [J]. *Satellite Applications*, 2016 (7): 45–47.

[9] Wang Jingquan. 'The Status and Development of the Commercial Remote-sensing Satellite Market' [J]. *Satellite Applications*, 2012 (1): 49–53.

[10] He Huaying, Wang Haiyan, Hao Xuetao, et al. 'Discussion on the current status and development of commercial remote-sensing satellite applications' [J]. *Satellite Applications*, 2016 (1): 68–71.

[11] Xu Liping. 'The Current Status and Thinking about the Development of the Commercial Remote-sensing Market' [J]. *Satellite Applications*, 2016 (1): 64–67.

[12] Yu Xiao, Liu Chang. 'The new sub-trend in foreign commercial aerospace from the perspective of recent industrial integration' [J]. *Satellite Applications*, 2017 (6): 24–29.

[13] Li Shuai, Hou Yukui, Man Yiyun, et al. 'The development of foreign commercial remote-sensing satellites' [J]. *Satellite Applications*, 2016 (3): 61–65.

[14] Lin Laixing, Zhang Xiaolin. 'Welcoming the 'orbital revolution'—the rapid development of micro satellites' [J]. *Spacecraft Engineering*, 2016 (2): 97–105.

CHAPTER 6

NEW OPPORTUNITIES FOR THE SATELLITE CONSTELLATION BUSINESS

In the 1990s, the development of satellite constellations for communication and network services reached a climax. However, with the rapid development of ground communication systems, many deployed constellations went bankrupt or fell into trouble due to losses. Since 2010, with the maturity of mobile communication technology, a consistent reduction of operating costs, and the development of various applications, the construction of a wide-ranging, economical, and practical Internet that integrates voice, data, and video has become an important infrastructure for promoting global economic growth. In this context, so called 'satellite-based Internet constellations' have attracted widespread attention again, offering new business opportunities.

As seen by the decline, revival, and recovery of satellite-based Internet constellations in the past 30 years, the question of how to deal with the relationship between constellations and ground communications is the key to success. The current development of emerging satellite-based Internet constellations is not a simple repetition of history, but the result of the comprehensive development of the market and technology. Looking to the future, opportunities and risks coexist. The key to success still lies with the market and users.

1. New Opportunities for the Space-based Internet

The geosynchronous equatorial orbit (GEO) satellite is located in the equatorial plane, and its daily cycle is consistent with the Earth's rotation period. Only three GEO satellites can achieve almost global coverage, which can be widely used in the fields of communication and television transmission. However, these satellites have high orbital altitude, higher path losses, longer signal delays, and difficulty in achieving full global coverage.

A satellite constellation is a group of satellites that are arranged in the Earth's medium and low orbit according to certain rules. Since the 1990s, due to the development of mobile communication and the Internet, the constellation of non-GEO communication satellites has developed rapidly, sparking the first climax for the space-based Internet. However, with the impact of the rapid development of ground-based mobile communication systems, satellite constellation communication has not been widely used due to its high cost.

Since the introduction of mobile communication technology and the Internet into public life in the 1990s, an urgent need has arisen for 'ubiquitous' communications and networks. This has pushed the ground-based mobile communication network from 2G and 3G to 4G in the past 20 years, and it is rapidly evolving to 5G. Although the ground-based mobile communication network now covers more than 80% of the Earth's population, it does not cover places such as remote areas, airplanes, and ships at sea. According to the United Nations, more than 3 billion people worldwide are still not connected to the Internet.

The need for faster, stronger, and wider use of the Internet has driven the development of satellite-based Internet technology. In fact, when the ground networks were still in the early 2G phase, Motorola introduced the first global mobile space-based communication network – the Iridium system – and began to deliver a global mobile communications service in 1998. Due to its high cost and slow construction process, it was quickly surpassed by the more cost-effective ground-based mobile communication scheme, and finally went bankrupt. However, the Iridium system and its contemporary Globalstar and OrbComm satellite communications constellations in low-Earth orbit have developed both technologically and commercially, creating a second generation of constellations that has laid the foundation for the development of today's space-based Internet.

1.1 The development of the space-based Internet

In the past 10 years, satellite communication and ground-based mobile communication have put an end to the industry-side competition for frequency, auguring more comprehensive and coordinated development of space-Earth integration.

The space-based Internet refers to a new type of network that provides broadband Internet services for ground and air terminals by using various air platforms located over the Earth. The space-based Internet has two main infrastructure models: the first provides signals to the ground through high-throughput satellite constellations at different attitudes and orbits ranging from 200–36,000 km. In recent years, satellite communications are moving toward high-throughput and broadband. Companies such as Intelsat, Viasat, and HughesNet in the US, and Eutelsat in Europe, are deploying high-throughput satellites in geostationary orbit equipped with Ka-band on a large scale. In low- and medium-Earth orbit, in addition to the three constellations (including Iridium), the European O3b constellation has been put into operation. The second model provides signals to the ground through near-space platforms, such as high-altitude balloons, airships, and unmanned aircraft, with a height of 20–100 km. In recent years, near-space platform technologies have matured. Google has proposed Google Loon balloons, while Facebook is working on high-altitude solar drones, and both have started to promote the globalization of Wi-Fi coverage.

Overall, high-throughput satellites are far more mature in their technology than near-space platforms, and will soon be available for large-scale commercial services. For example, the American Viasat uses Ka-band high-throughput satellites to provide free on-board Internet services to a number of American airlines. Progress in the satellite field will be discussed below.

High-throughput Satellites (HTS) are a new type of satellite that can provide up to ten times more capacity than traditional satellites under the same orbit and spectrum conditions through point beam, shaped beam, and frequency multiplexing technology.

Ka-band satellite technology began to develop in early 2005, and has become the focus of broadband satellite development due to its high frequency, rich frequency resources, and small terminal size. After two generations of HTS evolution, it has formed a complete industrial chain around the world. Satellite operators can build a large-capacity Ka-band satellite in less than three years. System vendors have supported technologies such as wide carrier, beam switching, and network roaming. Terminal suppliers have experience in the design of Ka-band terminal and antenna, and the corresponding technology has already reached the level of scale application. For satellite frequency resources, the traditional L-band and Ku-band are already very crowded, while the Ka-band is relatively free to apply for frequency and orbital position. More than 1,000 satellite network applications have been filed at ITU. In terms of cost, according to the survey, the investment cost based on Ku-band per Gbit/s will exceed 200 million US dollars, while the same unit cost of two Ka-band satellites (ViaSat1 and Jupiter 1) launched successively in 2011 and 2012 in the United States, was reduced to 4 million US dollars, which is only about 2% of Ku-band communication satellites.

The global space infrastructure that supports the space-based Internet has upgraded rapidly in recent years. The space-based Internet, which is currently forming commercial service capabilities, generally adopts the mode of a satellite constellation. According to the classification of satellite orbit type, it can be divided into a low-Earth orbit (LEO) satellite constellation system, a middle-Earth orbit (MEO) satellite constellation system, and a high-Elliptical orbit (HEO) satellite constellation system. Regardless of the altitude of the orbital satellite constellation system, the overall trend is to rapidly evolve to an HTS satellite.

The HEO satellite constellation system is developing rapidly. Viasat Inc. planned to launch ViaSat2 – the next-generation broadband satellite with a total capacity of about 300 Gbit/s – from the Ariane 5 rocket in June 2017, and would deliver services for over 700,000 end users. Meanwhile, a triple satellite constellation program ViaSat3 has been announced, with the single satellite capacity of 1 Tbit/s, covering the whole world from 2019.

The MEO satellite constellation system is developing steadily. As the world's first MEO communications constellation, the 12-satellite O3b system has started to provide services to maritime vessels around the world. The name 'O3b' stands for 'Other 3 billion', as it aims to deliver network services to another 3 billion people on Earth, focusing on providing access and backbone transmission where ground communication is currently unavailable, known as 'air optical fiber'. After commercial services were put into use in 2014, it only took six months to reach the original revenue level of 100 million US dollars per year. The development prospect of satellite-based Internet constellations has been proved by market acceptance. The O3b constellation has been acquired by SES – the second largest satellite operator in the world. By 2018, the O3b constellation will have 20 satellites, with the capacity of 12 Gbit/s per satellite.

LEO satellites are developing rapidly. The OneWeb constellation system plans to develop an LEO satellite constellation consisting of 648 satellites working in the Ku-band, with single throughput greater than 6 Gbit/s per satellite and a total system throughput of around 5 Tbit/s. When completed, the system can provide an Internet broadband access capacity of 50 Mbit/s to users in remote areas. The company also recently announced a development program for the next phase, consisting of 2,882 satellites. In addition, SpaceX, which has achieved remarkable success in developing launch vehicles, has also released the STREAM Internet Constellation Program, and has declared its frequency and orbital position to the ITU. According to the declaration, it has 4,425 satellites with both the Ku-band and Ka-band, running on 43 orbital surfaces. In addition, many other companies have announced LEO or MEO broadband satellite constellation programs. Thus far, development has reached unprecedented heights.

1.2 The application of satellite-based Internet constellations

Generally speaking, the navigation constellation and military Earth observation constellation are funded by the government. In contrast, the communications and

Internet constellations are commercial programs for profit. In fact, satellite and ground-based communication are both components of the telecommunication industry, and the two are developing in competition, complementation, and cooperation. Considering the competitive and cooperative relationship between ground-based communication and satellites, the development phases of communications and Internet constellations can be divided into three stages. The first phase began in the late 1980s and went up to 2000, and was represented by Iridium system. It involved the construction of a space-based network, and delivering to end-users via satellite telephones or Internet terminals, in competition with ground-based telecom operators. The second phase lasted from 2000 to 2014, and was represented by the new generation of Iridium, Globalstar and Orbital Communications, which provide some capacity supplement and backup for telecom operators, as well as mobile communication services for end-users in extreme conditions such as maritime and aviation. They have a certain degree of competition with ground-based telecom operators, but are regarded as 'filling gaps' for ground-based communications on a limited scale. Since 2014, the third phase (represented by O3b) has provided trunk transmission and cellular backhaul services to users worldwide. While ground-based telecom operators are customers and partners of O3b, satellite networks become the complement of ground-based networks.

Satellite-based Internet constellations have great value in emergency rescue, homeland security, and overseas military presence. Their development in foreign countries is focused on the commercial sectors, pursuing the huge economic benefits brought by Internet access either directly or indirectly. However, satellite-based Internet constellations should not be ignored due to their public and military potential in addition to commercial use. China has a vast land terrain, and more than 3 million square kilometers of marine territory where civil aviation passengers have not yet been able to access the Internet. At the same time, most remote rural areas and Gobi Desert in the west are still blind zones for communication. It is impossible for existing means of communication to protect the interests of the ocean economy and military activity. A global satellite communication system is an important solution to bridge the digital divide in these areas, and a way to offer emergency rescue in China and abroad. At present, China has the largest number of Internet users in the world, and the Internet can be used for voice, video, data, and other applications. Therefore, it is imperative to build a large satellite-based Internet system that will be coordinated by government organizations and cooperated among enterprises, to satisfy China's core interests and bring important strategic value.

It should be noted that the LEO constellation currently provides powerful Positioning, Navigation, and Timing (PNT) services, and can be used to enhance services when satellite navigation systems are not available. The addition of LEO navigation signals has many benefits. Compared with the MEO Global Navigation Satellite System (GNSS) constellation, the LEO satellite constellation is closer to Earth, with less signal dissipation and higher signal strength when reaching the ground.

Iridium is an LEO system that offers global coverage for communications services. Comparing the LEO Iridium constellation composed of 66 satellites with the GPS constellation composed of 31 satellites in the MEO, the orbital altitude of the two constellations is very different (several Earth radii). As a result, the intensity of the Iridium signal reaching the ground is 300–2,400 times higher than the GNSS signal. As a result, Iridium signals are an attractive solution for PNT applications when GNSS signals are limited. To achieve the same coverage capability as GPS at the orbit altitude of Iridium satellite, the number of LEO constellation satellites would need to be another order of magnitude. Future LEO constellations will have coverage comparable to GPS system.

By the year 2020, satellite-based Internet constellations will have undergone major developments, with their total capacity reaching or exceeding 7 Tbit/s. Their development brings many business opportunities as well as huge risks.

2. The Rebirth of the Iridium System

A constellation of small satellites operating in LEO began to provide communication and Internet services in the 1990s. They included the Iridium system consisting of 66 satellites, the Globalstar system consisting of 48 satellites, and the OrCam system consisting of 28 satellites dedicated to cargo monitoring. Not long after the three constellations were deployed, they all declared bankruptcy due to losses. Another larger Telegenic constellation of 840 satellites (later adjusted to 288 satellites) for broadband Internet access service was canceled without implementation. Many lessons have been learned from the failure of the previous generation of communication satellite constellations.

2.1 Why the Iridium system did not succeed

In 1987, Raymond J. Leopold, Ken Peterson, and Bary Bertiger – engineers from Motorola's satellite communications division – jointly proposed the idea of developing a constellation of LEO satellites 780 kilometers above the Earth to provide wireless communications at any point in the world. These satellites would be connected by wireless links to form a space communication network, which would act as a cellular base station in space. The wireless signal would be transmitted not through the cellular ground-based station but through the communication link between one satellite and another. In 1991, Motorola decided to establish a mobile communication network consisting of 77 LEO satellites (the number of satellites was later reduced to 66), and named this constellation after the 77th element 'Iridium' on the Periodic Table. Following up were the Globalstar and OrCam systems, which consisted of 48 satellites, but they did not have inter-satellite links. Iridium began commercial operations in November 1998. Due to the small number of users, it operated at a loss. On August 13, 1999, Iridium filed for bankruptcy with the courts. Globalstar and OrCam also declared bankruptcy in the years that followed.

In 2000, the author published an article called Iridium apocalypse, which analyzed the reasons for Iridium's failure as follows:

'First of all, there have been major errors in the prediction and foresight of technological development. Generally, it takes a very long period of time for large-scale aerospace programs to complete their development cycles. For example, the Iridium constellation took more than 10 years from the proposal to going into operation, while the renowned International Space Station (ISS) took 15 years to complete. Therefore, in the demonstration of large-scale aerospace system engineering, it is necessary to scientifically predict (or forecast) and foresee technological development and demand at least 10 years in advance. Undoubtedly, this is a very difficult task. The Iridium satellite system failed to correctly predict the future development trend of mobile communication technology, which was its biggest failure. At the same time, Iridium executives were not able to respond quickly to changes in technology and the market. Of course, it is not easy to change the scheme of large aerospace programs such as Iridium. In the proposal presented for the Iridium satellite system, according to the development level of the telecommunication industry at that time, it would have had a lot of business users. What was unexpected was that great changes had taken place in the mobile communication market after 10 years of development, and mobile ground-based communication technology achieved a major leap from analog to digital with relatively low prices. Chinese mobile users alone have exceeded 100 million people. The development of mobile ground-based communication technology has been faster than that of satellite mobile communication technology, based on the fact that the development of mobile ground-based communication technology depends on the development of electronic technology, which can be expressed by Moore's law. The development of satellite mobile communication technology depends not only on electronic technology, but also on the development of aerospace technologies and integrated technologies such as machinery, power, electronics, and materials. Obviously, in the past 10 years, the development of space technology has lagged far behind the development of electronic technology.

Second, there have been mistakes in market positioning. Satellite communication has the advantage of high space location, but it also has the disadvantage of higher cost. In general, the satellite communication market should be positioned to complement the deficiency of the ground-based system. It can be seen from the trends of the current and future global mobile communication market that, with the coverage of the mobile ground-based communication network getting wider and wider, the goal of using a mobile phone to travel around the world and access the Internet anytime and anywhere will no longer be a distant dream. As a result, the number of users of global satellite mobile phones will be significantly reduced. The market for satellite mobile phones will be positioned in special fields, and as a complement to ground-based communications.

Finally, the technology is complex and the cost is too high. The design scheme of the Iridium system shows the impact of performance-oriented concepts rooted in aerospace engineering in the past. That means considering more advanced techniques in the design scheme but less cost control in the program. In fact, the factors that dominate the competition are lower call charges and high communication quality. The cost of the call depends on the investment, lifespan, and maintenance of space and ground-based infrastructure.

Compared with other LEO systems, the Iridium satellite system had a higher call cost, partly because it had more satellites, and partly due to the complexity of satellites caused by its on-board processing ability and inter-satellite data link technology. For example, its inter-satellite data links included the communication links of two satellites behind and in front of the same orbital plane, and two satellites at the front and rear of the adjacent orbital plane. The transmission rate reached 25 Mbit/s by the Ka-band, and the technology was very complex. Therefore, compared with the Globalstar system, although the orbit height was low, the weight of the Iridium satellite was much heavier. This led to an increase the cost of satellite development and production, and also raised the cost of launching. Obviously, the on-board processing ability can improve the signal-to-noise ratio, and inter-satellite data links can reduce signal transmission time and improve signal quality, but it greatly increases costs.

Since the development of the Iridium satellite system, many people had advocated strengthening the space-based functions of spacecraft. They emphasized the development and application of on-board processing ability, on-board autonomy, and inter-satellite link technology. In fact, the technology a space system should adopt will be selected according to its mission requirements, based on the principle of the high-efficiency cost ratio, and based on the optimization of a large, integrated geospatial system in order to determine which subsystems should be placed in space or on the ground. For a satellite communication system, the ground-based system is inexpensive, reliable, and easier to integrate with the ground-based communication system.

For the LEO satellite constellation, the development and production costs of satellites account for one third to half of the total system cost. The total cost of the rocket (including development costs, product fees, and launch fees) is much higher than the total satellite cost (including development costs, product fees, and measurement and control fees). As the number of satellites increases, the costs is higher. In general, China should adopt the mode of one rocket launching two satellites or multiple satellites in order to share the production and launching cost of the rocket. At the same time, for reducing the total system price by cutting off the lifetime cost of the constellation, the launching cost of the space carrier vehicles must be reduced first. Only when the cost of space transportation is reduced by one or two levels can a market-oriented space enterprise like Iridium succeed.'

In hindsight, it seems that the above analysis is correct. It was only because the Globalstar system did not go bankrupt at that time that there was still an expectation for it. In fact, although the Globalstar system did not use inter-satellite link technology, the other aspects of the failure are similar to Iridium system, meaning that they were inevitable.

The story of the Iridium system provides much food for thought. As one American aerospace expert said: 'They provided the wrong product to the wrong market at the wrong time.'

2.2 Rebirth of the Iridium system

The Iridium satellite system was the product of high technology, and was the first large-scale LEO mobile communication satellite system put into use in the world. It opened a new era of global personal communication, and enabled people to communicate seamlessly in any place on Earth where the sky can be seen. The influential American magazine *Popular Science* awarded the 1998 Electronic Technology Award to Iridium. However, its meteoric fall happened within the space of a decade, from its appearance at the roadshow in 1990 to its bankruptcy in 2000. Nobody could have predicted what happened next.

According to US media reports, Iridium was auctioned for 10 million US dollars per satellite at a hearing held in the afternoon of March 17, 2000 at the New York Federal Bankruptcy Court. Despite the extremely low price, no one seemed to care. Instead, the judge ordered Iridium to take its 66 satellites out of orbit and let them burn up after entering the Earth's atmosphere. After consulting with several US government agencies, Iridium said it would take eight to nine months to get all of satellites into lower orbits, and estimated that it would take one to two years to burn them all up at a cost of 30 million to 50 million US dollars. The company was given some breathing space by the Pentagon, which was optimistic about its military use and did not want to see it disappear.

2.3 Dan Colussy – a business legend

Pan American Airline president Dan Colussy saw business opportunities in the bankruptcy of Iridium. He began to intervene in the transaction, and succeeded in saving Iridium. This is a classic case in the history of modern American business.

First, Colussy spent a lot of time persuading the US government to approve his involvement in the operation of the Iridium satellites. He was then awarded a five-year contract valued at 36 million US dollars per year from the Department of Defense (DoD), representing about half of Iridium's annual operating costs, to provide unlimited communications services to 20,000 government employees. Furthermore, Colussy persuaded the Department of Defense to compensate private companies for losses suffered in safeguarding national security interests by some of its little-known powers. After the matter was resolved, Motorola transferred its ownership and sponsors to Colussy at the end of 2000, and sold Iridium's 66 in-orbit satellites and 12 in-orbit standby satellites (some of the 88 previously launched satellites had failed to enter orbit).

Colussy put a price of 25 million US dollars to creditors (including Motorola) for the satellites, and secured a 2 billion US dollar liability insurance policy from Lloyd's Insurance Company in London to prevent the satellites from falling out of orbit and hitting surface objects.

The order from the Department of Defense (DoD) provided initial funding and customer support for Iridium's rebirth. Support also came from companies associated with Iridium – Motorola for its new handset, and Boeing for satellite maintenance and operations. Since then, Iridium's debt has been written off and the company has risen from the ashes.

Iridium began to provide Internet connection services in 2001 and an SMS service for 40 cents per message in 2003. Colussy began to focus on driving down the cost and decreasing the charge rate of a call from 7–13 dollars to 1.5 dollars per minute with a service charge of 10 cents per minute. At the same time, the price of mobile terminals dropped from 3,000 to 1,300 dollars, while pure data transceivers dropped from 800 to 250 dollars. In July 2004, Iridium began to turn a profit. In November 2008, it launched Iridium 9555 – a compact, light, and highly sensitive new satellite phone.

2.4 The military value of Iridium satellites

Although Colussy revitalized Iridium's assets, the real hope for recovery came from the September 11 attacks. Satellite phones provided critical communication links for the rescue effort. The terrorist attacks destroyed many ground communication networks in Manhattan, causing mobile phone users and ordinary telephone users to switch to satellite phones. Iridium donated and sold more than 1,000 of its satellite telephones to relief workers and institutions.

Satellite phones have come to the fore as government agencies and commercial companies have seriously re-evaluated their communications needs in emergencies. The sale and use of Iridium mobile phones are increasing rapidly. Since September 11, Iridium's total call time has increased 25 percent, and its total call costs are four times than what they were before.

The subsequent war in Iraq proved the military value of Iridium. According to Pentagon officials, 80% of the US military's satellite communication capacity during the Iraq War was provided by commercial satellite mobile communications such as Iridium. Its biggest customer is the Pentagon, which accounts for 21% of its profit. In 2004, on the first anniversary of the Iraq War, the US military assessed the coalition's combat plans, strategies, and technical equipment in Iraq, saying that officers and soldiers who had fought there recognized the crucial role of Iridium mobile phones. They stated that it would have been very dangerous to stop and set up huge antennas when confronting the enemy, so satellite mobile communications were necessary in maneuvering warfare. Since then, Iridium telephones have been widely used in the combat zones of Afghanistan and Iraq. The military application of Iridium also illustrates the necessity of civil-military integration of space technology.

Iridium satellite telephones were also used in the rescue work after the 2008 Wenchuan Earthquake in Sichuan Province, confirming once again the important role of satellite mobile communication when dealing with emergencies in marginal areas.

There are several reasons for Iridium's renewed success. The first is that the United States has a bankruptcy protection system, and the second is that the US Department of Defense noted in the military value of Iridium. Of course, Dan Colussy's commercial operation must not be overlooked.

An asset-heavy satellite company that filed for Chapter 11 bankruptcy protection, Iridium was taken over by Colussy with a 36 million US dollar deposit from the US Department of Defense. It must have cost more than 5 billion US dollars, although there are now conflicting accounts of how much Iridium cost to build. Globalstar, which cost the same, was acquired by Thermo Capital for 43 million US dollars. The fate of OrCam is similar. Finally, the three new satellite constellation companies have all managed to shift billions of dollars' worth of debt and return to profitability.

2.5 Upgrading the three LEO constellations

Since 2010, all three LEO communication constellations have been upgraded. Globalstar was the first to launch a second-generation satellite program due to the failure of the S-band transmitter of the first-generation satellite, which affected two-way data transceiver application. From 2010 to 2013, 24 second-generation satellites were launched, each weighing 700 kg. In addition to being compatible with fixed and mobile networks on the ground, the second-generation Globalstar has added satellite-based Wi-Fi services that allow users to access the Internet directly from their smartphones.

After failing to launch six replacement satellites in 2008, OrCam completed the launch of six second-generation satellites (OG2) in 2014–2015, with the two generations of OrbComm's satellites serving together in orbit. The OG2 satellite is equipped with an Automatic Identification System (AIS) to provide tracking for ships, and is a leader in satellite-based Machine-to-Machine (M2M) IoT applications.

The most complex and expensive constellation – Iridium-NEXT – successfully launched its first 10 satellites, each weighing 860kg, on January 15, 2017 using SpaceX's Falcon 9 rocket. Due to the complexity of the next-generation Iridium system and SpaceX's repeated delays in launching, it was not expected to be fully deployed and operational until 2018. The Iridium-NEXT system includes many new features in addition to its traditional capabilities. Each satellite currently has a payload of 50 kg to install an Automatic Dependent Surveillance-Broadcast (ADS-B) receiver from Airecon to provide near-real-time, high-precision aircraft position monitoring with global coverage. In the future, Iridium-NEXT will increase its payload margin, allowing service providers who want global coverage and inter-satellite links to complete their global constellation with the Iridium system instead of building their own satellite constellation.

So far, all three LEO communication systems have been reborn after periods of prosperity and bankruptcy because they all reduced the cost of satellite development and launch, enhanced the features of satellite constellations, and attracted new users. According to the revenue summary of the three companies given by the Federal Aviation Administration (FAA), their incomes for 2015 was as follows: Globalstar had the figure, with 9.49 million US dollars; OrCam came second with 178 million, and Iridium was the highest with 411 million. Clearly, Iridium and Globalstar have entered a plateau. OrCam has experienced rapid growth in recent years, especially in 2014–2015, and its revenue has increased by almost 70%.

Judging by the second-generation constellations of the three companies, they have clearly made full use of the advantages of their existing systems in the segmentation fields based on their existing users, shifting from competition to cooperation and supplementation with the ground-based communication network. For example, Iridium no longer competes directly with ground-based communication networks in terms of market positioning. Instead, it targets users not only in military fields, but also professional users in remote areas, such as offshore oil rigs, mining, construction, disaster relief, and off-country trips. As the cost of the system has reduced significantly, the cost of voice and data usage has reached a price level close to that of ground-based communications. Iridium has also made efforts to upgrade its satellite system in order to approach the capability of the ground-based system, reduce the size and weight of the satellite terminal, improve the speed of data service, and make itself competitive in specific application scenarios.

The successful upgrading of these three companies' constellations has encouraged more companies into the satellite-based Internet constellation market, representing a new climax for the satellite-based Internet constellation.

It has been more than 30 years since the introduction of the world's first satellite-based Internet constellation program in the late 1980s. Summarizing the failure, revival, and rebirth of each constellation, it can be concluded that the development of a commercial Internet constellation is not just related to technical issues. The key is learning how to deal with the relationships among various communication means on the ground in market competition. Although it has the feature of wide coverage, satellite-based Internet has the disadvantages of technical capability and service level compared with ground-based communication systems. Looking ahead, only satellite systems that are fully compatible with ground-based systems will succeed in the market. Integrated geospatial systems will be the way to success for satellite-based Internet constellations.

3. Greg Wyler's Constellation

Smart phones and e-commerce have profoundly changed consumer habits and the market environment. The rapid development of commercial aerospace has greatly reduced the cost of satellite development and launch, and has provided conditions for building the new-generation satellite-based Internet constellation. Greg Wyler, an American satellite practitioner, has played an important role in this development.

3.1 The Success of O3b

Wyler started in the American IT industry, and made his fortune by producing computer hardware. At the end of 2002, Wyler discovered by chance that the least developed countries such as Rwanda had no information services at all, so he decided to go there and do something about it. He founded Terracom, a telecommunication company whose business plan was to bring cell phones and Internet services to Rwandans. He managed the company from the USA, but made frequent trips to Africa. Workers – including Wyler, on occasion – dug trenches to lay hundreds of miles of fiber-optic cable, and set up Africa's first 3G cellular network. Terracom acquired the state-owned RwandaTel in 2005. This experience led Wyler to realize that hard-laid optical cables can only solve communication problems in an African country. To realize global interconnection, satellites are required.

In 2007, Wyler founded O3b with investments from SES Global, Google, cable operator Liberty Global, and HSBC, aiming to provide online services to 'the other 3 billion' people on the planet. The O3b constellation is the only satellite-based Internet constellation that has been successful since its inception.

The O3b constellation is deployed in mid-Earth orbit (MEO), and comprises 12 satellites, three of which are backup stars. Six more are planned for the future. Each satellite weighs 700 kilograms, and all satellites in the O3b constellation are manufactured by Thales Alenia Space. All of the satellites were successfully launched in 2013 and 2014, and the constellation operates in the Earth's equatorial orbit at the attitude of 8,062 kilometers, covering the area within latitude 45°North/South. O3b's satellites operate in the Ka-band, and each is equipped with 12 steerable spot beam antennas, allowing seamless handover among them. There is no data-link between the satellites, and all routing exchanges are carried out by ground gateway, which is then connected to the ground-based communication network. All communications between users should be relayed through the ground gateway

The service provided by the O3b constellation is similar to the traditional transponder lease service, but the current transponder is not fixed. It has to switch among different satellites according to their movement, and the orientation of each beam is adjustable.

As the O3b satellites work in MEO and use spot beam antennas, compared with the traditional telecommunication satellite working in the geostationary Earth orbit (GEO), they offer low latency, high antenna gain, and small signal loss. However, the O3b constellation cannot achieve seamless coverage; it can only adopt the hot spot coverage mode, letting the user's spot beams of each satellite point to respective target zones in different service areas, in order to improve the utilization rate of resources.

The O3b constellation does not provide services directly to individual consumers, but works through telecommunications and Internet operators. The system currently provides long-distance calling and Internet services for Asia, Africa, South America, and Oceania, and is particularly suitable for providing broadband access for cruise ships.

When O3b was first established, the market was still skeptical about the constellation system. However, since the system began to provide commercial services in 2014, it took only six months to reach the expected revenue level of 100 million US dollars, which was recognized by the market. The success of the O3b constellation system was firstly due to the development concept of cooperation with the ground system. From the beginning, there was no plan to compete with ground-based telecommunications. Instead, the telecom operators were regarded as their customers, serving the islands and large vessels that could not be covered by the ground-based communication facilities. They soon became a supplement to ground-based means of communication. Secondly, the O3b constellation system uses MEO and covers only the area between latitude 45° North/South, thus reducing the number of satellites required to 12, which greatly lowers the system input cost. Finally, O3b is a true broadband satellite system, and its data transmission rate greatly exceeds the Iridium and Globalstar systems. Although the system capacity cannot be compared with that of ground-based communication systems, even less than the GEO broadband satellite, it can already meet basic network requirements in areas where ground facilities cannot be reached.

3.2 The advantages of OneWeb

The success of the O3b constellation has inspired many investors to enter this field. WorldVu – the predecessor of OneWeb – was founded in 2014 by former O3b founder Greg Wyler. Wyler was head of the satellite-based Internet program at Google before leaving to start WorldVu.

The current financing situation indicates that the market is optimistic about the future development of the OneWeb constellation. On June 25, 2015, OneWeb raised the first round of 500 million US dollars of financing. The deployment, operation, and service of the OneWeb constellation depends on its capital partners. Satellite manufacturing is undertaken in collaboration with Airbus; satellite launches are carried out by Arianespace and Virgin Galaxy; Qualcomm is responsible for the design and construction of the air interface and dual-mode terminals; Hughes is responsible for the design of terminals and for distribution with Coca-Cola; Intelsat is responsible for the construction of ground-based gateway stations; India's Bharti Enterprises and Totalplay Telecommunications are responsible for distribution and services in the Indian and Mexican markets; and Rockwell Collins and Honeywell are responsible for aviation terminal services, and share users and services with Intelsat. On December 19, 2016, Korean-Japanese Masayoshi Son decided to invest 1 billion US dollars in OneWeb.

The OneWeb constellation will provide global telecommunications operators with fiber-optic quality Internet access services. The initial constellation of the first-generation OneWeb includes a total of 882 satellites (648 satellites and 234 backups), and most of its bandwidth has been booked. Each satellite weighs 150 kg, and a technique called Progressive Pitch is used to rotate the satellite slightly to avoid interfering with the Ku-band satellite in geostationary orbit.

In February 2017, OneWeb submitted an application to the US Federal Communications Commission to build a larger constellation. It plans to manufacture, launch, and operate an additional 2,000 small in-orbit satellites. These in-orbit satellites will reach approximately 2,882 other satellites. The constellations will operate in LEO and MEO via the Ku-band and Ka-band.

On June 15, 2016, OneWeb and Airbus Group announced the establishment of a joint venture satellite manufacturing plant to design and manufacture a total of 900 small satellites at one time. The new manufacturing plant will take Airbus for reference in terms of industrialization, standardization, and automation in aircraft production to produce these small satellites. The cost per small satellite will be reduced to about 500,000 US dollars. The 85 million dollar plant, located in Florida's Discovery Park, was inaugurated on March 17, 2017. It will be able to produce three satellites a day.

On June 25, 2015, OneWeb announced the largest ever commercial rocket acquisition, including 21 Soyuz launch orders from Arianespace and 39 launches from Virgin Galactic's LauncherOne. On March 9, 2017, OneWeb signed several contracts with Blue Origin for five separate launches, with about 400 small satellites to be launched from 2021. Blue Origin will use more capable, reusable New Glen heavy rockets for these launches.

So far, at least 20 large LEO satellite constellation programs have been proposed worldwide, among which the OneWeb constellation is the most advanced. The frequency, funds, chip technology, satellite manufacturing, launching, and marketing have been clearly defined, so OneWeb constellation have obvious advantages.

Firstly, they have registered for the occupancy of the frequency band and orbital position. At present, except for the OneWeb constellation that has confirmed the Ku-band, other companies have not yet obtained frequency resources. If a company cannot obtain frequency band and orbital position resources as quickly as possible, the development progress of its satellite-based Internet constellation program will be delayed.

As for the successful cooperation between O3b and some telecom operators, instead of selling dedicated satellite terminals like Iridium phones and competing for customers with fast-growing smartphones and tablets, OneWeb-based users can continue to access satellite networks with existing smartphones and tablets.

To establish a sales channel that combines tradition and innovation, OneWeb has chosen Hughes Networks – a company that is well versed in broadband satellite operation and related ground equipment sales – and uses its perfect sales channels and rich sales experience to focus on traditional industry users. It has also chosen to cooperate with Coca-Cola to make use of the multinational's sales outlets all over the world in order to get closer to the mass consumer market. At the same time, Coca-Cola retail stores can serve as network users of the OneWeb constellation.

On March 4, 2019, OneWeb finally launched the first six of the 650 global Internet satellite clusters planned for the first phase.

3.3 The emergence of SpaceX

Following OneWeb, SpaceX announced its LEO broadband satellite constellation program with 4,425 satellites, and Boeing announced its 2,956 V-band satellite constellation program. The original antenna equipment supplier Kymeta proposed a constellation of LEO satellites (LeoSat) with 80 to 140 satellites. Some high-Earth orbital telecom satellite operators such as Viasat and Telesat have also announced plans for LEO/MEO broadband satellite constellations.

Three of the most representative solutions are OneWeb, SpaceX, and LeoSat. What the three satellite-based Internet constellation programs have in common are that they all chose small satellites with a quality of less than 300 kg, and all broadband Internet service. The differences are that OneWeb is focused on individual consumers and communities, while LeoSat is primarily focused on business operation users. OneWeb and SpaceX have a larger number of satellites, while LeoSat uses a powerful satellite platform with a high-throughput satellite (HTS) capacity of estimated 108 satellites.

The most ambitious of these plans is SpaceX's proposed STEAM satellite-based Internet constellation. The STEAM satellite-based Internet project reflects SpaceX's expectation for the future development of satellite manufacturing, hoping to gain more profit from satellite manufacturing than launching, and accumulating experience and capability in satellite development.

On November 10, 2014, Elon Musk announced that SpaceX was collaborating with WorldVu satellite, founded by Greg Wyler, to develop an LEO constellation composed of 700 satellites that would provide Internet access services to the whole world. However, the cooperation between SpaceX and WorldVu was not successful. Greg Wyler later renamed WorldVu to OneWeb. In January 2015, Musk proposed that SpaceX would develop its own STEAM constellation. It planned to build an Internet constellation of more than 4,000 small satellites to provide Internet access services worldwide. Musk believes that this project could even be used for future Martian migration. It is his hope that when humans land on Mars in the future, this technology will provide Internet services there.

Instead of buying satellites from traditional manufacturers, SpaceX is focused more on manufacturing. It hopes to master the upstream of the satellite industry chain and expand from manufacturing and launching to services. As a result, the company will need more financing. According to SpaceX's estimates, a total of 10 billion to 15 billion US dollars will be needed to start a satellite manufacturing plant and build more than 4,000 LEO satellites. It has already received 1 billion US dollars in funding from Google and Fidelity Investment in the previous round.

Compared to OneWeb, the biggest problem SpaceX faces today is that it is difficult to obtain frequency band and orbital position resources. SpaceX has reported frequency bands and orbital positions to the International Telecommunication Union (ITU) through the Norwegian government. According to the report, there are 4,257 satellites operating on 43 orbital planes using the Ku-band and Ka-band. SpaceX also reported to the ITU for additional six to eight experimental satellites in the Ku-band through the FCC. The first satellites – MicroSat-1a and MicroSat-1b – are expected to have a life span of one year, and operate in circular orbits with an orbital inclination of 86.6° and an orbital altitude of 625 km.

On May 3, 2017, at a hearing on broadband services in the US Senate, SpaceX released more details about its satellite access plan, and expressed hope that the US government would modify its corresponding regulatory system. SpaceX had previously applied to the Federal Communications Commission (FCC) to launch 4,425 satellites for Internet service. By the end of 2017, SpaceX will carry out satellite-based Internet technology testing, which will continue until the beginning of 2018. If the tests go smoothly, SpaceX plans to launch satellites in batches from 2019 to 2024, after which the system will be put into commercial use.

More radically, SpaceX launched a V-band LEO program in March 2017, raising the number of satellites in the constellation to an astonishing 7,518.

In fact, investment in the construction of the satellite-based Internet constellation is huge, and the prospects of the market are still very uncertain. Therefore, its future development will be full of both opportunities and risks. Certainly, the larger the constellation, the greater the risk. This has not escaped the attention of the media.

4.   Coexisting Opportunities and Risks

Looking to the future, the space-based Internet will be fully integrated with the fast-approaching 5G ground-based internet, and the space-Earth integrated Internet network will become part of everyday life.

4.1 Opportunities for the satellite-based Internet constellation

Traditional telecommunication satellite operators focus on purchasing satellites and providing capacity. Their profit model is simple, their income sources are fixed, and their market and development space are very limited. The emergence of the satellite-based Internet provides a new development idea for traditional telecommunication satellite operators. Most of these new satellite-based Internet companies have the industry background of the Internet, and their focus will be on consumer services and data. Satellites are regarded more as a medium for enterprises to realize their core competence. In building satellite-based Internet constellations, they can grasp a large amount of user data, analyze user needs from massive data, and develop end-user-oriented services to increase profits. Since the industrial chain is no longer divided by manufacturing, launching, operating, and service in the traditional sense, it can promote the cooperation and integration of upstream and downstream industries, which will certainly create a new business model and bring rich industrial income.

At present, the ground-based mobile Internet has been widely applied in every industrial field, and its application has reached an advanced stage. However, there are many occasions when the ground-based Internet still cannot provide good coverage and application. The demand for networking on these occasions is more urgent than the areas that the Internet can cover. For example, settings such as the outdoors, fields, aircraft, and the sea require access to and application of broadband Internet. According to a survey, more than 73% of travelers would like to surf the Internet when they are outdoors, while nearly 100% would like to do so when they are in forests, snowy mountains, or deserts. This is possible with the space-based Internet.

The good news is that at present, the space-based Internet based on satellite constellations is highly mature, and some satellites even have the basis of scale application. For example, the GEO and MEO constellations have exceeded expectations in aviation, navigation, and other markets after being put into commercial operation. Another example is the LEO constellation. With the joining of several Internet and IT giants, the number of individual users has exploded. There are reasons to believe that the next three to five years will be a golden age for space-based Internet satellite constellations. It can be predicted that in the next five to 10 years, the space-based LEO Internet platform will accelerate its maturity.

First, the application of smartphones has brought new profit models to the satellite-based Internet. In the past, the satellite-based Internet constellation was only used as means of uncovering areas and supplementing ground communication networks, and the user scale was very limited. It was difficult to meet investors' return expectations simply by calls and internet charges. Nowadays, the rapid development of smartphones and e-commerce has changed users' consumption habits and altered the market environment. Consumers have become accustomed to paying for content services. Communication and network services are no longer the only ways for operators to profit. In the future, once the satellite-based Internet is successfully established, it will be the equivalent of a huge data portal that can provide a variety of services, thus forming a huge commercial market.

Second, the aviation industry will continue to develop new applications. In the early morning of March 8, 2014, the disappearance of Malaysian Airlines flight MH370 proved that in the Internet era, anything from automobiles and mobile phones to household appliances can be connected to the Internet. However, the data systems of commercial aircraft are rarely connected to any network, showing that the civil aviation information system is very much behind the times. The disappearance of MH370 indicates the urgency of developing a satellite-based air management system to replace the current ground-based radar system. To this end, FlightAware and Aireon have begun to cooperate in the development of a Global Beacon system to provide satellite-based ADS-B services through the Iridium constellation. These are scheduled to be operational in 2018, and will meet the requirements of one-minute interval flight tracking in 2021.

Accessing the Internet on flights is an urgent requirement for the vast majority of passengers, and an important market for the satellite-based Internet. In 2005, Airbus launched the world's first in-flight Wi-Fi network, which uses the Globalstar system to connect to the Internet at high altitudes. The United States currently has the best in-flight Wi-Fi coverage, and most of its airlines are already offering Wi-Fi service to passengers. American Airlines offers Wi-Fi on 78% of its flights.

In-air Wi-Fi and In-Flight Entertainment Connectivity (IFEC) have been deployed by airlines in recent years. In the face of fierce market competition, airlines regard IFEC services as a new profit growth point, hoping to increase users and expand new business models. At present, the demand for on-board Internet has evolved to a bandwidth of more than 50 Mbit/s per aircraft. Otherwise, it is difficult to ensure that passengers on board can enjoy the same or similar mobile Internet access as on the ground. For international flights over remote areas, it is difficult to ensure uninterrupted service with existing communication satellites, which creates business opportunities for the development of satellite-based Internet constellations.

Finally, the development of commercial space travel has led to significant cost reductions. The failure of satellite-based Internet in the 1990s was due to the high cost of constellation input, which in turn was due to the high cost of producing and launching satellites. New launch companies like SpaceX have already cut launch costs by 50%, and advances in reusable rocket technology are expected to bring the cost down to around 20 to 30% of what it used to be. In addition, OneWeb and SpaceX are incorporating the mass manufacturing experience of aircraft and missiles into satellite development, in the hope that this will bring the cost down to less than 500,000 US dollars per satellite. With the development of small satellite technology, increasing numbers of satellite projects are beginning to use non-aerospace-grade off-the-shelf commercial components, and some widely recognized official or non-official standard and modular interface designs have been formed for mass manufacturing and rapid launch. With the application of 3D printing technology, small satellites or satellite components can be directly produced, thus reducing the development cost and cycle.

4.2 The risks of satellite-based Internet constellations

Opportunities and risks coexist in the development of satellite-based Internet constellations. The key to future success depends on whether the service costs can be reduced to prevail in the market and attract users.

First, compared with the various technical means of providing broadband Internet services, satellite-based Internet constellations are facing competition from ground-based and other satellite systems. As a compromise, the time delay – a key indicator of network performance – of the LEO constellation is close to the ground-based optical fiber and cellular communication, which meets the requirements of interactive communication applications such as VOIP (Voice over Internet Protocol), video conferencing, and online games. It is much more effective than the GEO broadband satellite and MEO constellation. In terms of user cost, the LEO constellation is close to the GEO broadband satellite, but there is still a gap between the LEO constellation and means of ground-based communication. Although the system capacity of the LEO constellation greatly exceeds the GEO broadband satellite, the centralized capacity provided by the GEO broadband satellite in terms of the spot beam still has more advantages. Compared with global coverage and the ability to bridge the digital divide, the LEO constellation has obvious advantages.

Ground-based mobile operators are now making significant efforts to promote the development of 5G. It is difficult to predict whether these constellations will fall behind ground-based communication again when they are completed. At the same time, other Internet companies (including Google) are also trying out various high-altitude communication system solutions, including releasing thousands of giant balloons in the atmosphere to provide a network for the 'Loon' program in remote areas, while Facebook is now trying to provide high-altitude Internet services through the integration of stationary GEO satellites, drones, and laser communications.

The American satellite and space industry consulting firm Northern Sky Research (NSR) roughly estimates the future revenue of Internet satellite companies by using the debt ratio as an economic indicator. At 50 US dollars a month per user, NSR calculates that each company that is building the satellite-based Internet constellations of the future will need at least 10 million users to make a profit. This is a very difficult goal to achieve. For example, Viasat had only 1 million broadband users in 2016, after its first Ka-band GEO broadband satellite was launched in 2011. Even with payload and network content revenue, these emerging satellite-based Internet companies will find it difficult to recover costs only from network business revenue. Although the global market for unconnected and connected users is still very large (an estimated 471 million), most of these users are in low-income countries, where there is still considerable uncertainty in terms of market development.

Secondly, the coordination of frequency resources is very difficult. OneWeb was the first to obtain the right to use the Ku-band from the ITU, but OneWeb does not have ownership of this band, and other competing companies can still apply. For this reason, other firms must first co-ordinate with OneWeb and with geostationary satellite operators that have applied for spectrum. Because of the competitive relationship between them, coordinating frequency resources is very difficult.

In addition to this, there is also the issue of landing rights. If services are to be provided globally, frequency resource coordination with the governments of various countries is also required. It is relatively easy to obtain frequency authorization in regions and countries such as Europe, the United States. and Brazil, but in other countries a lot of work is needed.

Finally, the harsh space environment of the LEO must be withstood. For example, the OneWeb constellation plans to use an electric propulsion system to reduce the mass of satellite launching, cut the cost of launching, and increase the number of launches. The rocket first sends the satellite into an orbit of 500 kilometers, and then sends it to a predetermined orbit position through an electric propulsion system. In the process of changing a satellite's orbit, it is necessary to cross the high-energy proton radiation belt frequently, which will have a major impact on the normal operation of the satellite's electronic system. Therefore, anti-radiation capability is required.

LEO satellites are affected by space radiation, and also run the risk of collision with other satellites, micro meteors, and orbital debris. In 2009, Russia's Cosmos 2251 spacecraft collided with the Iridium 33 satellite at an attitude of 790 kilometers, and some 600 pieces of debris were detected. With the increasing amount of space debris, the probability of it colliding with LEO satellites is very high. Currently there are more than 20,000 pieces of debris above ten centimeters, and more than 500,000 pieces larger than one centimeter. Space debris is particularly dense in the orbit commonly used by the satellite-based Internet constellation. The impact of large debris on spacecraft can be fatal, and small debris may also cause damage. The cumulative effect of debris below a millimeter in size may also affect the work of satellites. This is a major issue in the construction of satellite-based Internet constellations.

In order to reduce the probability of collisions in satellite-based Internet constellations, the orbital center of each constellation is required to be separated by a certain distance. The positioning accuracy of each satellite should be improved, and in-orbit propulsion can be carried out to implement anti-collision operation. A decommissioned satellite must be off-orbit within a certain period of time. These requirements will inevitably increase the cost of building and operating constellations, and need to be coordinated by the appropriate international agencies.

The emerging satellite-based Internet constellation has only been in development for around three years since the end of 2014, and its system design, commercial channels, and profit model are still immature. However, in terms of future system size, market opportunities, and system capabilities, it has broad prospects for development, so a lot of capital, technology, and partners have been rapidly assembled in a short time. While at present, there is still a way to go to achieve the stable operation of the system and the mature development of the industry, considering the tremendous changes and industrial value offered by the Internet, there will be tremendous business opportunities in the fields of investment and financing channels, system research and development, and product services in the future. These are the future development directions of commercial telecommunication satellite industry. Therefore, the development of new satellite-based Internet constellations abroad must forge ahead with tracking and research.

References

[1] Huang Zhicheng. *The Fourth Wave of Aerospace Science, Technology, and Society* [M]. Guangzhou: Guangdong Education Press, 2007.

[2] Huang Zhicheng. 'New Opportunities for Satellite-based Internet Constellations' [J]. *Space Exploration*, 2017 (7): 24–32.

[3] Huang Zhicheng. 'The Iridium Apocalypse' [J]. *Electronic Outlook and Decisions*, 2000 (3): 11–18.

[4] Huang Zhicheng. 'The Progress of Satellite Communication in Non-geostationary Earth Orbit' [J]. *Electronic Outlook and Decisions*, 1998 (3): 44–46.

[5] Huang Zhicheng. 'Small and miniature modern satellites' [J]. *Science & Technology Review*, 1998 (2): 10–13.

[6] Huang Zhicheng. 'New Opportunities for Satellite-based Internet Constellations' [J]. Space Exploration, 2017 (7): 30–33.

[7] An Hui. 'The Rebirth of Iridium' [J]. *Space Exploration*, 2017 (7): 34–35.

[8] An Hui. 'The Success of the O3b Constellation' [J]. *Space Exploration*, 2017 (7): 36.

[9] An Hui. 'The Advantages of the 'One-network constellation''[J]. *Space Exploration*, 2017 (7): 37.

[10] Liu Yue, Liao Chunfa. 'The Development of Foreign Satellite-based Internet Constellations [J]. *Science & Technology Review*, 2016, 34 (7): 139–148.

[11] Liu Yue. 'The influence of emerging satellite-based Internet constellations on the development of traditional satellite manufacturing' [J]. *International Space*, 2016 (11): 51–58.

[12] Du Feng, Li Guangxia, Zhu Hongpeng, et al. 'An overview of the development of next-generation small satellite constellation communication systems' [J]. *Satellite Applications*, 2016 (5): 14–19.

[13] Zhang Youzhi, Wang Zhenhua, Zhang Xinxin. 'The Development Status and Analysis of Europe's O3b Constellation System' [J]. *International Space*, 2017 (3): 29–31.

[14] Blue Sky Wing. 'The development of commercial aerospace from two generations of non-geostationary communication satellite constellation systems' [J]. *International Space*, 2017 (3): 7–18.

CHAPTER 7

COMPETITION FOR LAUNCH VEHICLES

In the field of launch vehicles, SpaceX successfully developed the Falcon 9 rocket as well as the recovery technology for the first-stage rocket. In doing so, it has reduced the launch cost, built a foundation for the development of heavy rockets with greater carrying capacity, and prepared for the future development of transport systems for exploring the Moon and Mars. Meanwhile, Blue Origin is following in SpaceX's footsteps by developing a large launch vehicle. The success of these two commercial aerospace companies has forced traditional rocket launch companies to develop a new generation of large low-cost rockets, and has encouraged more commercial outlets to enter the market for launching small satellites, forming a situation of fierce competition.

1. The Falcon 9 Rocket and its Recovery Technology

The Falcon 9 is a two-stage rocket. Both the first and second stages use a Merlin 1D engine fueled by liquid oxygen/kerosene. The first stage uses nine Merlin 1D engines, and the second stage employs a Merlin 1D vacuum-type engine with an increased nozzle expansion ratio. In other words, the main engine of the rocket uses only one type of engine. This design concept greatly reduces the rocket's need for an engine system. The Falcon 9 has a total length of 70 meters and a diameter of 3.66 meters. It has a take-off thrust of 6,806 KN and a take-off weight of 541 tons

1.1 The Merlin Rocket Engine

The main task for a company that began with the development of launch vehicles is to find a rocket engine that meets demands. When Elon Musk established SpaceX in 2002, he considered buying Russian rocket engines, as it is very difficult to develop a liquid rocket engine. After repeated failed negotiations, he was forced to develop one by himself.

To this end, Musk hired several engineers with extensive experience of engine development, including Tom Mueller – chief designer of TRW's TR-106 rocket engine. The engineers has used two key technologies in the Apollo Moon Landing Program for the descent engine of the Lunar Module: a pintle injector and a low-cost ablative cooling carbon fiber composite nozzle, which helped to develop a Merlin 1A rocket engine powered by liquid oxygen/kerosene for propellant with a gas-generator cycle mode.

In 2006, the Merlin 1A engine was installed and put into use on the Falcon 1 rocket. So far, Merlin engines have been in service for 11 years. During this period, SpaceX has made consistent improvements, deriving the Merlin 1B, Merlin 1C, Merlin 1C Vacuum version, Merlin 1D, and Merlin 1D Vacuum version. The latter two have been used as the main engine of the Falcon 9 rocket, and will continue to be used in the Falcon Heavy rocket when it is launched.

During the evolution of Merlin engines, their nozzle structure, cooling mode, and turbo pump have been improved, leading to enhancements in their thrust and specific impulse. When thrust increases were completed at the end of 2016, the sea-level thrust of the latest Merlin 1D engine rose from 34.6 tons to 86.2 tons (an increase of 150%) and that of the vacuum version rose to 93.2 tons. The specific impulse of the sea level increased to 282 s, and that of the vacuum version to 311 s. Thanks to improvements to Merlin engines and the maximum 40% throttling capacity, the Falcon 9 became the first rocket capable of vertical reverse thrust recovery, and its LEO carrying capacity also increased from 9 tons to 22.8 tons.

The Falcon 9 follows the design concept of the general engine. The first stage is powered by nine Merlin 1D engines, and it is very difficult to guarantee the reliability of so many engines at the same time. SpaceX has solved this problem by means of power redundancy. All nine first-stage engines on the Falcon 9 have redundant capabilities that allow them to keep the payload in proper orbit by extending the working time of the remaining engines if one fails. In fact, during the CRS-1 mission in October 2012, when an engine failed, the rocket was still able to put the Dragon spacecraft into orbit. This redundancy design has also been used on Saturn 1B rockets in the United States.

The Falcon 9 uses a large number of lightweight aluminum-lithium alloys and composite structures to help reduce the dry weight of the full rocket. It also uses the common bottom tank design of two large tanks that store liquid oxygen and kerosene (which the Saturn 5 rocket used) to lower the dry weight of the full rocket and reduce the height of the rocket. This design can save one of the tank bottoms, but has a high requirement for the insulation layer. The Falcon 9 rocket reduced manufacturing costs by using mature processing technology from the US aerospace industry – general-purpose processing equipment that can be used to manufacture 3.66-meter-diameter rockets, and the extensive use of off-the-shelf redundant commercial electronic components. As a result, the Falcon 9v1.0 was the lowest-cost rocket in the world.

1.2 The example of DC-X

Not satisfied with the success of the Falcon 9v1.0 rocket, Elon Musk put his mind to developing reusable rockets to reduce launch costs and get to Mars. To this end, SpaceX concentrated a large amount of resources (about 1 billion US dollars) to test reusable rocket technology. At the same time, it worked on upgrading the existing Falcon 9 rocket to improve its performance and meet the requirements of reuse.

In 2011, SpaceX built a first-stage Falcon 9v1.0 rocket called the Grasshopper, which was equipped with four steel landing legs, and repeated eight tests in increasingly high jumps to accumulate experience of vertical rocket landing. The original Grasshopper was taken out of service in 2013, and the larger Grasshopper 2 (based on the Falcon 9v1.1 first-stage rocket) was put into use. Unfortunately, Grasshopper 2 was destroyed during its fifth flight due to an anomaly that was detected in the vehicle, and the autonomous flight termination system terminated the mission.

In both SpaceX's Grasshopper rocket and Blue Origin's New Shepard rocket, the reusable concept comes from the single-stage vertical take-off and landing rocket DC-X (Delta Clipper Experimental) from the 1990s. On 18 August 1993, Dr. William A. Gaubatz of McDonnell Douglas (now incorporated into Boeing) trialed a one-third scaled test rocket DC-X/A with a flight time of 59 seconds at the White Sands Missile Range (WSMR) in New Mexico, USA. Two more test flights were conducted in September of the same year. The final flight reached 150 meters, after which it slowed down and hung in the air like a helicopter. The rocket then tilted slightly and flew about 150 meters laterally. It then started to land tail down, rapidly stretching out its retractable legs with disc-shaped feet and slowing down with flames from its four small engines for a smooth landing. The whole flight lasted 66 seconds.

The DC-X program was developed as a technology demonstration project for the US Department of Defense. After 1993, the project was transferred to NASA and DARPA and adopted advanced materials to build the rocket's structure. In 1996, DC-XA made several more test flights, setting a record for flying 142 seconds and reaching an altitude of 3,140 meters. Unfortunately, during the last test flight, the DC-XA crashed to the ground and was destroyed. Although the project ultimately failed to gain continued support from NASA and the US military, it demonstrated the possibility of single-stage VTVL (Vertical Take-off and Vertical Landing) rockets being used to transport conventional payloads into LEO. Its multiple flight tests also accumulated technologies related to reusable VTVL rockets, such as the ability to start and stop engines precisely and allow a safe return after an engine failure.

1.3 The first reuse of a rocket

In order to realize recoverable soft landing after launch, a rocket must overcome two technical key points: engine technology that can adjust the thrust in a large range, and multi-constraint guidance and control technology in the re-entry phase.

The propellant of the Merlin 1D engine of the Falcon 9 rocket is fed into a combustion chamber via a turbo pump. At the same time, the turbo pump provides high-pressure fluid for the hydraulic actuator, which then recycles into the low-pressure fuel inlet. This eliminates the need for a separate hydraulic drive system, and satisfies the requirement to adjust the thrust value over a wide range.

The Falcon 9v1.0 rocket made five successful flights between 2010 and 2013 before it was replaced by the Falcon 9v1.1 with a 60% increase in take-off thrust. Although its diameter has not changed, it has a new engine compartment structure, longer propellant tanks, higher thrust, and a more efficient Merlin 1D engine. The Falcon 9v1.1 rocket made its first flight on September 29, 2013. Since then, it has succeeded in 14 of 15 launches, but all three landing tests failed.

Meanwhile, the Falcon 9v1.1 evolved into the Falcon 9v1.2, equipped with legs for landing on offshore platforms. On December 22, 2015, the Falcon 9v1.2 launched for the first time, while its first-stage rocket landed safely near the launch pad. The Falcon 9v1.2 uses ultra-low temperature propellant near melting point to increase its propellant density, allowing the rocket to carry more propellant in the same volume of tanks.

The first stage of the Falcon 9v1.2 has nine engine nozzles and four grid fins. After the first stage was separated from the second stage, the former performed a procedure turn of 3-engine boost back, and performed a 180 deg flip in the sky. The nine main engines of the first stage faced the ground and kept three engines burning to perform a re-entry deceleration, adjusting the attitude of the first stage. When the rocket came close to the ground, the four grid fins on the top of the first stage were deployed to stabilize the attitude. At this point, the main engine fired again, using a thrust slightly below the weight of the rocket to slow it down. The first stage of the rocket used a guidance law with terminal angle, speed, and position constraints to approach the ground landing field and implement a soft landing to achieve the first stage of rocket recovery.

Some of SpaceX's recovered rockets landed on the ground, while others came to rest on floating platforms offshore. Even though SpaceX can recover the first-stage rocket from the ground, it also needs to be recovered from offshore platforms. Due to the influence of ocean waves, it is difficult to ensure that the inclination angle and position of the rocket is in the ideal state on the offshore platform. This means that recovery offshore is much more difficult than on the ground. Whether or not to make a recovery at sea depends on the trade-off between the rocket's carrying capacity and recovery performance. Reusable rockets need to carry more fuel than expendable rockets, which means increasing their weight and reducing their ability to carry payloads. For a recovery on the ground, the rocket needs to fly a certain distance towards the designated landing site after re-entry. In a recovery at sea, the offshore floating platform can be towed to the first-stage rocket landing area, and no longer needs to fly a certain long lateral range. According to the existing conditions, if the Falcon 9v1.2 rocket is recovered on land, it will lose about 40% of its carrying capacity. If the recovery site is changed to the sea, only 16% of the payload capacity will be lost. The platform for the Falcon 9v1.2 rocket in the Atlantic Ocean, is 91.1 meters long and 52. 05 meters wide, and is equipped with four 300-horsepower engines to maintain stability. Its name is OCISLY (Of Course I Still Love You).

On March 30, 2017, a recovered first stage of a previously-flown Falcon 9v1.2 rocket launched a Luxembourg-based SES-10 commercial communications satellite into orbit after several successful sea recovery attempts. As the first successful launch of a satellite by a recovered rocket, this was another milestone in the history of manned space travel. The rocket's grid fins were the last product to use aluminum alloy. Subsequent grid fins have been made from titanium alloy, the better to deal with the possible impact of the rocket's engine exhaust.

By the end of June 2017, the Falcon 9 rocket had launched 37 times, including nine times in 2017 and twice by the recovered first stage. The first stage was recovered 13 times, eight of which were from the offshore platform.

On April 9, 2016 at 04:00 Beijing time, SpaceX's recoverable rocket made its first perfect landing at sea on a drone ship. By April 12, 2019, SpaceX had successfully recovered 36 rockets, including 24 at sea and 12 on land.

SpaceX is currently developing the Falcon Heavy rocket. Its center core is exactly the same as first stage of the Falcon 9v1.2 rocket, and its first-stage booster is also the same Falcon 9v1.2 rocket bundled on both sides. It will rely on 27 Merlin 1D engines plus one upper stage to put the payload into orbit. Its performance is roughly the same as that of the SLS heavy launch vehicle being developed by NASA. In July 2017, Elon Musk publicly stated that the risk of the Falcon Heavy rocket would inevitably rise as the number of engines increased, and the recovery of the rocket would become more difficult.

On February 6, 2018, SpaceX successfully launched the Falcon Heavy rocket with the strongest carrying capacity in the world. It succeeded in launching a Tesla Roadster into space, and also recovered two side boosters. As the most powerful super rocket in the world, the Falcon Heavy can put a payload of 63.8 tons in lower-Earth orbit (LEO) or 16 tons onto Mars. On April 12, 2019, the Falcon Heavy successfully launched an Arabsat-6A communication satellite of the Arabsat League in Saudi Arabia. A few minutes after the launch, the rocket's three boosters were separated and returned to Earth. The boosters on both sides of the rocket landed on SpaceX's concrete platform off the Florida coast, while the larger central booster landed on the SpaceX drone recovery ship in the Atlantic Ocean.

Elon Musk's ultimate goal is to send humans to settle on Mars, and the vehicle for this goal is SpaceX's newly developed Starship. On April 3, 2019, the prototype of the Starship completed a tethered hop ignition test.

SpaceX plans to use the Starship spacecraft to take people to the Moon and Mars. The stainless steel vehicle will be mounted on top of a large booster called the Super Heavy and launched into deep space. Like SpaceX's current Falcon 9 rocket, the Starship's boosters can land on Earth and be reused after the launch. The Starship/Super Heavy rocket combination should be able to launch up to 100,000 kg into LEO, making it one of the most powerful rockets ever launched. The Super Heavy booster will use one of SpaceX's new-generation massive Raptor engines. In February 2019, SpaceX tested a full-scale version of this engine.

2. The Raptor Engine and the Interstellar Transportation System

On September 28, 2016, Elon Musk gave a keynote speech titled 'Making Humans a Multiplanetary Species' at the 67th International Astronautical Congress in Mexico. He also launched the Interstellar Transportation System (ITS) for human colonization of Mars, proposing a manned landing on Mars by 2025. The longer-term plan is to realize human migration to the Red Planet.

At present, SpaceX's Mars program still has major uncertainties in terms of economy and technology. In July 2017, Musk publicly stated that after a period of research, the ITS solution would make major changes.

2.1 The ITS solution

Since man landed on the Moon in 1969, Mars has been the next target for human space exploration. The questions are, who will be the first to land on it, in what way, and when? Colonizing Mars and making humans into a multiplanetary species have always been Elon Musk's visions since the establishment of SpaceX in 2002. In his aforementioned speech, he pointed out that Mars is the most suitable planet for colonization. To reach it over such a long distance, it is crucial to dramatically reduce the cost of launch and operation. In order to solve this problem, he proposed several key measures.

The first is to build a fully reusable rocket. Musk uses the example of the Boeing 737, which costs 90 million US dollars and carries 180 passengers. If it were expendable, tickets from Los Angeles to Las Vegas would cost as much as 500,000 US dollars per person. Now, because of repeated use, tickets only cost 43 US dollar. If the present expendable launch vehicle can be fully reused, the cost can also be greatly reduced.

The second is to realize orbital refueling. SpaceX's launches will be divided into ships (for passengers) and tankers (for cargo or fuel), and a tanker will be used to refuel ships with propellant. This will reduce the size of launch vehicles. If orbital refueling is not adopted, a launch vehicle with a three-stage structure needs to be developed, and the size and weight of the rocket should be at least 5-10 times larger than the current two-stage version.

The third is to produce propellant on the surface of Mars that returns to Earth. Astronauts will use Martian water and carbon dioxide to produce methane and oxygen on the surface of Mars to provide enough propellant for the re-entry spacecraft.

Fourthly, in order to reduce development costs and improve reliability, ITS will continue to adopt the method of bundling several Falcon 9 rocket engines of the same type.

In summary, the development of rocket engines to meet the above requirements has become the top priority. The result is the Raptor engine – a full-flow, staged combustion engine powered by cryogenic methane and liquid oxygen (LOX).

2.2 The deficiencies of the Merlin engine

Before SpaceX proposed the ITS solution to Mars, it conceived the Merlin 2 engine as a way of developing the Falcon heavy launch vehicle. It would adopt the design of Merlin 1 and scale up proportionally, with the same fuel combination and ability to throttle from 70% to 100%. It would have a thrust of 7,560 kN at sea level, and a specific impulse of 285s. The vacuum version of Merlin 2 would provide a thrust of 8,540 kN, and a specific impulse up to 321s. However, faced with the goal of landing on Mars by ITS from SpaceX, the Merlin engine still has many shortcomings.

Firstly, it is not a fully reusable engine. A major problem affecting the reuse of engines is the carbon deposit formed after combustion. This deposit increases the maintenance cost of the engine and affects the number of times it can be reused. The Merlin engine uses LOX/kerosene as fuel, which leaves severe carbon deposits.

Secondly, the thrust of the Merlin 2 engine is still not adequate. The take-off thrust required by the ITS rocket mentioned above is 13,303 tons, which is unlikely to be solved by increasing the number of Merlin 2 engines.

Thirdly, as the specific impulse of Merlin 2 engine is still low, the efficiency of an upper-stage engine will also be low.

Finally, it is very difficult to produce kerosene on Mars.

Based on the above considerations, SpaceX abandoned the idea of developing the Merlin 2 engine. Instead, it decided to develop a Raptor engine that uses LOX/methane as fuel and employs the full-flow staged combustion cycle.

2.3 Why choose LOX/methane as propellant?

The LOX hydrocarbon rocket engine is significant because of its non-toxicity, high average density, and relatively high performance. As well as the wide application of LOX/kerosene rocket engines, since 1980, many countries and regions (including the United States, Russia, Europe, Japan, and China) have carried out research into the characteristics of LOX/methane rocket engines, such as high specific impulse, low cost, low carbon deposition, non-coking cooling, and suitability for reuse.

Similar to hydrogen and oxygen, methane is a cryogenic propellant, and its mainte-nance conditions are the same as those of liquid oxygen (LOX). The physical properties of hydrogen, oxygen, methane, kerosene, and unsymmetrical dimethyl hydrazine (UDMH) are shown in table 7-1.

Table 7-1   Comparison of the Physical Properties of Propellants

Properties	Hydrogen	Oxygen	Methane	Kerosene	UDMH
Molecular Weight	2.016	32	16.043	167	60.78
Boiling Point (K)	20.38	90.2	112	466–547	336
Freezing Point (K)	13.95	54.4	90.65	226	216
Critical Temperature (K)	33.23	154.8	190	658	521
Critical Pressure (MPa)	1.32	5.08	4.6	1.82	5.26
Density (liquid) ($kg \cdot m^{-3}$)	70	1140	422	836	791
Specific Heat at Constant Pressure ($J \cdot kg^{-1} \cdot K^{-1}$)	15,000	1,700	3,480	1,980	2,734
Coking Temperature (K)	None	None	950	560	None

Table 7.1 shows that:

1) The density of liquid methane is half that of kerosene, and about six times that of liquid hydrogen. Therefore, liquid methane storage tanks are much lighter than liquid hydrogen storage tanks.

2) The coking limit temperature of kerosene is the lowest and methane is the highest in hydrocarbon fuels. The coking limit temperature of kerosene is 560 K and methane is 950 K. Therefore, methane does not coke at lower temperatures. Electrical heat transfer experiments were carried out in 1980 in the United States to study the coking characteristics of hydrocarbon fuels. The results show that methane can work normally at a wall temperature of 773 K, and there is no coking when the sulfur content of methane is less than 1×10^{-6}.

   Hydrocarbon fuel gas generally has carbon deposits. In the United States, carbon deposition studies on hydrocarbon fuels have been carried out with a mixing ratio of 0.2 to 0.6 and a combustion chamber pressure of 50 to 120 MPa. The results show that there is no carbon deposition in the mixing ratio range of methane. Methane molecules contain only one carbon atom, and it is difficult to form long-chain hydrocarbons after pyrolysis, so carbon deposition is not easy at high temperatures. The experiment shows that there is no obvious carbon deposit in the LOX/methane rich combustion products in the wide combustion temperature range of 673 to 1,173 K.

   A test conducted in 1986 by Aerojet on the compatibility of hydrocarbon fuels with combustion chamber walls showed that when the sulfur content of methane was less than 1×10^{-6}, there was almost no corrosion to the inner wall of the copper alloy.

3) Methane has a higher specific heat. Its specific heat at constant pressure is lower than that of hydrogen, but higher than that of kerosene and other propellants, and it does not coke. Therefore, methane makes a good regeneration coolant.

4) The boiling point of methane is 112 K – less than 22°C from the boiling point of liquid oxygen, 90.2 K. Therefore, a common bottom tank can be used to simplify the tank structure.

   Specific impulse is an important index for evaluating the performance of various rocket engines. The theoretical specific impulse of various types of non-toxic propellant engines will vary with the mixing ratio. Among all hydrocarbons, the specific impulse of LOX/methane is the highest. The highest theoretical specific impulse of a LOX/methane engine is lower than that of a LOX/LH2 (liquid hydrogen) engine, but higher than that of a LOX/kerosene engine. In the Russian LOX/kerosene RD-180 rocket engine, when the mixing ratio is 2.72, its specific impulse at sea level is 311.3s, while specific impulse of its vacuum version is 337.6s. According to calculations by the German Aerospace Institute, if it is changed to a LOX/methane engine, its specific impulse at sea level can be increased to 322.5s; the vacuum version can be increased to 348.3s.

Liquid methane is almost as safe to use as liquid hydrogen. Methane is not toxic; the percentage of its explosive volume is 5%–15%, and its automatic ignition temperature is 813 K. Methane is flammable, but it has a smaller molecular weight and is lighter than air. Like hydrogen, any leakage of methane can immediately rise and escape into the atmosphere. As a result, methane is still safe to use according to safety guidelines.

Methane resources are abundant. Liquid methane comes from liquefied natural gas (LNG) and solid gas hydrate (flammable ice). Natural gas hydrate, which is almost pure methane, is widely available around the world, and its reserves are at least twice as large as those of conventional fossil fuels. There are also plenty of methane resources on Mars and Saturn. Methane is cheaper than 1/70 of liquid hydrogen and 1/3 of kerosene.

In Elon Musk's aforementioned keynote, the performance of the three propellant combinations (LOX/kerosene, LOX/LH2, LOX/methane) was compared as a graphic. The results show that LOX/methane is exceptional in terms of rocket size, propellant price, reusability, and its capacity to function as Mars production propellant. In propellant transportation, it is as good as LOX/kerosene but better than LOX/LH2.

In summary, the LOX/methane engine has the comprehensive advantages of a hydrogen/oxygen engine and an LOX/kerosene engine, and is one of the future development directions of the aerospace propulsion system.

2.4 The innovation of the Raptor engine

In Musk's keynote, the parameters officially released for the Raptor engine were 311 tons of sea-level thrust (3050 kN) and 357 tons of vacuum thrust (3,500 kN), using a precooled LOX/methane propellant combination. If the Raptor engine is put into use, it will become the fourth largest active liquid rocket engine in the world, after the Zenit rocket's RD-171 (7,904 kN), the Atlas 5 rocket's RD-180 (3,830 kN), and the Delta IV heavy rocket's RS-68A (3,137 kN, a little larger than the Raptor). The specific impulse of the Raptor engine is 334s at sea level and 382s at vacuum.

Two days before Musk made his speech, SpaceX had successfully conducted a scaled Raptor engine ground fire test in Texas, USA. The thrust of the tested engine was 100 tons.

Traditional large launch vehicles generally require their booster engines to have high density-specific impulse (product of density and specific impulse), so LOX/kerosene engines are usually chosen. At the same time, a high specific impulse is generally required for engines of the core or upper stage, which makes LOX/LH2 engines a natural choice. In order to develop ITS, SpaceX decided to use an LOX/methane engine, which is an innovative way of developing heavy launch vehicles.

In order to meet requirements, the Raptor engine must increase the specific impulse. As a result, it has abandoned the gas generator cycle and adopted a new full-flow staged combustion cycle. It features a dual pump and dual precombustion chamber design. One of the precombustion chambers features oxygen-enriched combustion, while the other offers methane-enriched combustion. The high-temperature oxygen-enriched gas and high-temperature methane-enriched gas are introduced into the main combustion chamber to generate the combustion of two stable airflows to achieve higher performance. The Raptor engine has a combustion chamber pressure of up to 300bar, which is the highest single-combustion thrust of any staged combustion cycle liquid rocket engine. Moreover, the engine has a throttling capacity of 20% to 100%. At the same time, this design poses some critical problems such as its complex structure, increased weight, and unverified combustion stability.

To this end, SpaceX is taking a number of measures to solve these key issues. For example, it plans to use 3D printing technology to process components of the Raptor engine that account for 40% of the total weight, develop high-strength high-temperature resistant materials, and develop new computational fluid dynamics (CFD) simulation software for the Raptor engine.

Since the high development cost of the rocket engine comes from the ground test, in the past 30 years, many scholars have devoted themselves to the application of the CFD method to simulate internal flow, but the results have not been satisfactory. SpaceX has reportedly developed a unique CFD simulation software, and adopted spectra analysis based on the wavelet transform method, to solve multidimensional turbulence simulation within a different space and time. If the CFD simulation software can be verified by the ground test of the engine, it is expected that the number of ground tests for the Raptor engine can be greatly reduced, meaning that development costs can be cut.

In summary, according to the information released by SpaceX about the ITS program, it has several unique bright spots in terms of innovation. However, whether the project succeeds depends on the financial guarantee, and whether the difficulties of the Raptor engine can be overcome on schedule. As for the progress of SpaceX, it will be difficult, but worth waiting for.

3.   Blue Origin's New Shepard and New Glenn Rockets

Jeff Bezos' space activities began with a single-stage, fully reusable, vertical take-off and landing (VTVL) rocket called the New Shepard. Unlike SpaceX, Blue Origin's goal in developing reusable rockets was first to sell ordinary people a 'ticket' to travel through space, and transport passengers 100 kilometers above the Earth's surface. So far, the rocket has completed six test flights, with the last five using the same rocket.

After Blue Origin was founded in 2000, Bezos hired some of the engineers who had been involved in the development of the DC-X. It is not difficult to see the influences of the DC-X in the shape of the New Shepard rocket and the use of LOX/LH2 engines.

3.1 The New Shepard Rocket

The New Shepard is a single-stage reusable VTVL rocket. It includes a crew capsule and an LOX/LH2 propulsion section. The propulsion section is equipped with a single BE-3 LOX/LH2 rocket engine. It employs the air extraction cycle to directly extract a small part of the gas from the combustion chamber to drive the twin-turbine pump, eliminating the gas generator and the pre-combustion chamber, simplifying the structure, and improving reliability. The rated thrust of the BE-3 engine is 49.896 tons, which can be adjusted to 9.072 tons. In order to control the attitude of the rocket and make a precise landing, the tail of the rocket's body is equipped with four wedge-shaped fins and eight expandable drag brakes. To slow down and landing, the rocket is equipped with three parachutes, a reverse thrust engine, and six landing legs.

The top of the rocket is equipped with a crew capsule for sightseeing, which is 15 cubic meters in size and is scheduled to carry six passengers. The capsule is separated from the propulsion section at the highest point after lift-off, and continues to climb to the sub-orbital space about 100 kilometers above the ground by means of inertia. Passengers will have about four minutes to experience weightlessness. The capsule then returns to the ground under gravity. It is decorated with white bulkheads as well as comfortable seats and cushions, meaning that passengers can admire the Earth from high altitudes.

In 2006, Blue Origin conducted an experimental flight by a scaled test rocket, climbing to an altitude of 87 meters and landing successfully. Another test rocket was used for a flight test, reaching an altitude of 14,000 meters, exceeding the DC-XA flight altitude record. At the same time, the flight Mach number of the rocket reached 1.2, but because the angle of attack (AOA) exceeded the limit, the rocket crashed. The accident greatly affected the development of the New Shepard rocket.

On April 25, 2014, the New Shepard rocket was launched from a test site in Western Texas. The rocket climbed to 93.5 km, and the crew capsule returned safely to land after separation. However, the propulsion section lost control due to a failure of the hydraulic system, resulting in a crash. After fixing the hydraulic system, Blue Origin achieved complete success in a flight test with another New Shepard rocket on November 23, 2015. The flight reached 100.5 km altitude, reaching Mach 3.72. The passenger capsule separated and returned to land safely. During the return of the propulsion section, the control system withstood a crosswind of up to 191.47 m/h at high altitude, meeting the design requirements and keeping the propulsion section stable. Under the action of drag brakes, the speed of the propelling section decreased to 622.68 m/h. At this time, the BE-3 engine ignited again, which reduced the speed of the propelling section to 7.08 m/h, and allowed for a safe, soft landing.

On January 22, 2016, Blue Origin successfully launched and recovered the same rocket it had recovered previously. The flight altitude was close to 102 kilometers, which was higher than that of the previous flight test. This test improved the software for the guidance, navigation, and control system, and allowed the rocket to control a more convenient landing position in the landing area after initially aiming at the center of the landing zone. This increased the margin for allowing deviation, and improved the ability of the rocket to withstand wind disturbances at low altitudes.

Since then, flight tests for the New Shepard rocket have entered the fast lane. Blue Origin conducted three successful recovery tests with the same rocket on April 2, June 17 and October 5, 2016. In the last test, a successful escape test of the crew capsule was also carried out.

In March 2017, Jeff Bezos stated that the New Shepard rocket would start manned test flights from 2017, and would travel to space with paid passengers in 2018.

Having spent 15 years on the development and construction of the New Shepard rocket, Bezos finally succeeded thanks to sheer perseverance. His success is a completely different story of entrepreneurship to Elon Musk's.

3.2 The BE-4 rocket engine

The key to developing inexpensive, reusable large launch vehicles is the engine. Obviously, it would be difficult for Blue Origin to reduce launch costs if it used the existing BE-3 LOX/LH2 engine, so Bezos proposed to develop the BE-4 rocket engine with LOX/methane as a propellant. It has a sea level thrust of 250 tons, which is greater than the 190 tons of the main engine of the space shuttle.

When Bezos talked about the idea of developing the BE-4 engine, he said: 'In principle, rocket engines are simple, but that's the last place they are ever simple.' Nonetheless, Blue Origin sought to make an engine that was not too complex nor required ultra-premium materials. The designers did not want to create a work of art that pushed the limits of engineering. Rather, they wanted a reliable workhorse that could be flown again and again, perhaps as many as 100 times as the company pushed the boundaries of reusable space travel.

In 2012, Blue Origin began to develop its own BE-4 engine. In 2014, the United Launch Alliance (ULA) decided that its next generation of Vulcan Rockets would use two BE-4 engines instead of the Russian RD-180 engine used by the Atlas 5 rocket, and planned to launch it for the first time in 2019.

The RD-180 is a super-high-performance LOX/kerosene engine, operating at extreme temperatures and pressures. Higher pressures translate into marginally more performance, but at a high cost of development time, money, and uncertain reusability. Bezos said: 'Our strategy is that we like to choose a medium-performing version of a high-performance architecture. However, the RD-180 is a high-performing version of a high-performance architecture. That's a very challenging engine to develop, and it really complexifies everything. With lower pressure, you can still get very high performance.

The BE-4 engine should cost about 30 to 40% less than the RD-180 engine. It should also, in theory, be more durable and capable of reuse without risking the failure of components due to the extreme pressures and temperatures inside it, as it is flown once and then discarded. The BE-4 will use the staged combustion cycle adopted by the RD-180 engine, which will be more efficient than other current US liquid hydrocarbon rocket engines. Because the BE-4 engine will be built by adding materials, its quality will be lighter, it will be more productive, and the cost will be lower.

To support the development of the BE-4 engine, Blue Origin has built a ground test facility specifically for liquefied methane engines, capable of testing engines with thrusts above 450 tons. The facility was put into use in May 2014. The BE-4 engine has been tested for various components. On March 6, 2017, Blue Origin assembled the first BE-4 engine, and it is estimated that a full-scale hotfire test will be conducted soon.

3.3 The New Glenn Rocket

Bezos announced in September 2016 that Blue Origin will develop a new rocket capable of putting people and payloads routinely into Earth's orbit. The new rocket is called New Glenn, and commemorates John Glenn – the first American astronaut to complete an orbit around the Earth. The launch capability of the rocket will exceed that of SpaceX's future Falcon Heavy and the ULA's Delta IV Heavy.

On March 7, 2017, Jeff Bezos disclosed new details about the New Glenn reusable launch vehicle at the Satellite 2017 conference, and announced that it had signed a contract with Eutelsat to launch commercial communications satellites using the rocket. On March 8, Blue Origin announced a new partnership with satellite-based Internet service provider OneWeb, signing five satellite launch contracts. On May 1, 2018,

Blue Origin successfully launched and recovered the New Shepard rocket, reaching a maximum altitude of around 107 km. This test was a new attempt by Blue Origin to send humans into suborbit.

There are two versions of the New Glenn launch vehicle, with two stages and three stages. The first-stage rocket is recovered and is expected to be reused 25 times. The two-stage rocket has a diameter of about seven meters and a height of about 82 meters. If the third stage is added, the height of the rocket will be increased to 95 meters. The New Glenn has a payload capacity of about 13 tons to geosynchronous transfer orbit, and about 45 tons to lower-Earth orbit (LEO). The first-stage rocket will use seven BE-4 engines, with a total take-off thrust of about 1,750 tons – far more than the 950 tons of the Falcon 9 rocket. The rocket's second stage uses an upper-stage version of a single BE-4 engine. The third stage will use an upper-stage version of the BE-3 engine.

The New Glenn rocket has four forward fins as aerodynamic control surfaces at its first stage. After it is separated, the first stage will adjust attitude and ignite again to slow down the speed of descent and deploy six landing gears before landing. It will eventually land vertically on an unmanned offshore platform. Achieving recovery after the launch of an in-orbit rocket will open up a whole new situation for Blue Origin.

The New Glenn rocket will be built at Blue Origin's manufacturing center, which is currently under construction in Cape Canaveral, Florida. Bezos has said that the New Glenn rocket is scheduled to lift off from Launch Complex 36 (LC-36) at Cape Canaveral by 2020. He has also revealed that Blue Origin is working on a future New Armstrong rocket. From the name, it may be related to the Moon or beyond.

Bezos named his launch vehicle after the American space pioneer John Glenn, expressing his commitment to space exploration. Bezos' trajectory in this field has both similarities and differences with Elon Musk's. Musk aims to colonize space and spread human civilization across multiple planets. Bezos believes that the Earth is the most suitable planet for human habitation, and space development should be used to enhance it. The two entrepreneurs are also different in the way they started their businesses. Musk is radical while Bezos is steady. It is thanks to the diversity of American space entrepreneurship that the US commercial aerospace industry is at the forefront of the world.

4.   New Rockets from Traditional Companies

The success of SpaceX and Blue Origin has forced traditional rocket launch companies to develop a new generation of large, low-cost rockets. Here are just two examples.

4.1 The ULA's Vulcan

The United Launch Alliance (ULA) was established in December 2006 as a 50-50 joint venture between Lockheed Martin and Boeing. It brings together the most successful and experienced teams in the field: Lockheed Martin's Atlas and Boeing's Delta team. The ULA's customers include the US Department of Defense, NASA, and other aerospace organizations and companies.

Delta is a versatile family of expendable launch systems that has provided space launch capabilities in the United States since 1960. More than 300 Delta rockets have been launched with a 95% success rate. Only the Delta IV Heavy rocket remains in use as of August 22, 2019. The Delta IV series is 63–72 meters in height, five meters in diameter, and 249.5–733.4 tons in take-off weight. Its LEO carrying capacity is 9.42–28.79 tons, and its GTO carrying capacity is 4.44–14.22 tons. The core stage of the Delta IV rocket uses RS-27 or RS-68A rocket engines with LOX/LH2 as propellants and a thrust of 3140 kNb. The second stage of the rocket core uses an RL10-B-2 engine with LOX/LH2 propellant, and the thrust is 110 kNb. The Delta IV rocket can choose to bundle two or four solid rocket boosters, using hydroxyl-terminated polybutadiene (HTPB) as propellant, with a thrust of 826.6 kNb. The ULA plans to phase out all Delta IV launch vehicles except the Delta IV Heavy as far as possible by 2018 in order to improve competitiveness. The first stage of the Delta IV Heavy rocket consists of three Delta IV core stages in parallel, which are used to launch confidential national security payloads. However, the frequency of missions is very low, being carried out every few years or so. The ULA is working with companies such as XCOR Aerospace and Aerojet Rocketdyne to develop options for upper-stage engines. The rocket will not be certified for launching national security payloads until at least 2022.

The Atlas V rocket contains the 400 series and 500 series, which are two-stage rockets. They are 58.3 meters in height and 3.81 meters in diameter, with a take-off weight of 334.5 tons. The core first stage of the Atlas V uses an RD-180 rocket engine with an initial thrust of 3,827 kN, while the Centaur's second stage uses an RL-10A or RL-10C engine producing 99.2 kN of thrust. The rocket can choose to bundle 1–5 boosters according to its mission. The booster length is 17 meters, the diameter is 1.6 meters, the weight is 46.697 tons, and the thrust is 1688.4 kN. The LEO carrying capacity of the Atlas V rocket is 9.8–18.81 tons, and the geosynchronous transfer orbit (GTO) carrying capacity is 4.75–8.9 tons.

In response to SpaceX's challenges, the ULA announced major changes in its personnel and corporate architecture in October 2014 as a way of reducing launch costs. On April 13, 2015, it announced a plan to develop a new Vulcan launch vehicle. The plan took a gradual approach, starting with the first stage of the new launch vehicle, then developing a second stage, and finally completing the design of an engine that can be reused after each launch mission. The purpose of the ULA's development of the Vulcan launch vehicle is to compete with SpaceX and to replace the Russian RD-180 as the main engine of the Atlas V rocket. After the Crimea crisis in 2014, US Congress banned the use of Russian-made rocket engines in US national security missions.

By schedule, the first stage of Vulcan rocket is now in the process of development. Its first choice is the BE-4 LOX/methane rocket engine that is being developed by Blue Origin. At the same time, the ULA is also cooperating with Aerojet Rocketdyne to develop the AR-1 LOX/kerosene rocket engine for use if BE-4 becomes unavailable.

The second stage of the Vulcan rocket will be developed on the basis of the upper stage of the Centaur and the fairing of the Atlas V rocket. The diameter of the fairing is between four and five meters. The Vulcan rocket will be bundled with up to six solid rocket boosters. Its carrying capacity will exceed that of the Atlas V, but will be lower than that of the Delta IV Heavy.

After the first successful recovery of the first stage of the Falcon 9 rocket, the ULA held a press conference on April 13, 2016, saying that it would adopt second-stage rocket technology that could be reused in orbit on the Vulcan rocket. The second-stage rocket, carrying a lot of fuel, could become a so-called gas station in space. The ULA had previously announced that the first-stage rocket engine of the Vulcan rocket could be recovered by parachute. The upper stage of the Vulcan rocket will have enough fuel to work in orbit for seven to eight days, allowing it to perform multiple missions in space, such as space rendezvous and docking. If it can refuel in space, its life can be extended even further. This capability will make complex space missions possible, such as launching spacecraft components into orbit separately and then assembling distributed payload missions in orbit.

The ULA has said that the development cost of the Vulcan launch vehicle is about 2 billion US dollars, including 1 billion for the development of the main engine. The ULA and its strategic partners will pay for the total development cost, but will not refuse government investment. On November 30, 2016, the ULA released a website for customizing rockets: www. rocketbuilder. com.

4.2 The Ariane 6 Rocket

The Ariane rocket series was developed by the European Space Agency (ESA) – an intergovernmental organization established in 1975 that comprises 11 European countries. It is an outstanding achievement in global space history, and a successful case of international cooperation in aerospace technology. Since the launch of the Ariane 1 rocket in 1979, ESA went on to develop four iterations. Each new rocket was innovated on the basis of inheritance, greatly improving its performance and reliability. With the success of the Ariane 4 rocket, it quickly occupied the space launch market that originally belonged to the United States. Developed in the late 20th century, the successful Ariane 5 rocket put the American launch market under pressure with its large carrying capacity and high reliability.

There are many reasons why the Ariane series achieved so much commercial success. In terms of forecasting, ESA foresaw the great potential of the space launch market through careful research into space development and application. In terms of management, adopting a national-based, multi-national cooperative research and development policy not only strengthened unified leadership, but also gave full play to the advantages of all nations. In terms of technology, ESA has always adopted the development policy of setting a low starting point, and improving and reducing risks. I terms of the overall goal, ESA has always emphasized high standards, low costs, high reliability, and adaptability. It adopts rather flexible policies in market competition, such as low charges for new rockets for the loads carried during the first few launches.

The most commonly used rocked in the series is the Ariane 5. It started development in January 1988 and took its first flight in June 1996. So far, the 5G, 5GS, 5ECA, and 5ES models have been developed and put into use. With 480–746 tons of take-off weight and 11,400–13,000 kN of take-off thrust, Ariane 5 can be used to transmit all kinds of satellites and spacecraft to GTO, sun-synchronous orbit, MEO, LEO, and fly-off orbit. It is a two-stage liquid rocket with LOX/LH2 propellant in the cryogenic main stage and nitrogen tetroxide, monomethyl hydrazine, or LOX/LH2 propellant in the cryogenic upper stage. Two large solid rocket boosters can be bundled around the cryogenic main stage. The Ariane 5GS rocket's cryogenic main stage uses the Vulcain 2 engine, and the upper stage is powered by a liquid engine with multiple ignition ability and normal-temperature propellant. The GTO carrying capacity of the 5GS rocket is 6.5 tons. The Ariane 5 ECA rocket uses the improved Vulcain 2 cryogenic propellant engine, and the upper stage uses an HM7B cryogenic engine, increasing the GTO carrying capacity to 10.5 tons. The Ariane 5ES rocket was used to launch the European Automatic Transfer Vehicle (ATV) into LEO with a capacity of 21 tons.

Faced with a challenge from SpaceX, ESA held a ministerial conference in Luxembourg on December 2, 2014. It decided to create a new family of Ariane 6 launch vehicles and give greater responsibilities to commercial aerospace companies under the new management system. In this context, Airbus and Safran created the joint venture Airbus Safran Launchers (ASL) in January 2015, and completed the final establishment on July 1, 2016. A year later, on July 1, 2017, ASL officially changed its name to ArianeGroup.

ESA decided that the development of the Ariane 6 rocket would be led by the industry. On July 16, 2015, nearly 4.2 billion euros' worth of contracts were approved, covering the design and manufacture of the next generation of the Ariane 6 rocket, the construction of its launch site, and the development of an upgraded version of the Vega small satellite launch vehicle. The rest will be financed by ASL and its industry partners.

The Ariane 6 rocket is a low-cost launch vehicle. Its core stage uses a Vulcain engine powered by LOX/LH2 propellant. The Vinci – a new cryogenic engine for the upper stage – is now under development. The Ariane 6 rocket has two models. Bundled with two solid boosters, which are the first stage of the Vega rocket that launches small satellites, the Ariane 62 rocket can launch five tons of payload into the GTO. The Ariane 64 rocket, equipped with four boosters, would be able to launch a commercial communication satellite weighing up to 10.5 tons into the GTO. The French Guiana Space Center is located at 5° north latitude, close to the Equator, and can take advantage of the Earth's rotation to improve the performance of rocket launches.

By forecasts, the Ariane 6 rocket can halve the cost of launching of heavy communication satellites compared to the current Ariane 5 rocket. Compared with the Falcon 9 rocket, it has double the payload and volume, and a price more than two times lower. However, the price for each commercial launch of the Falcon 9 rocket is 60 million US dollars. By the time the Ariane 6 rocket makes its first flight in 2020, SpaceX will have long been able to reuse the first stage of Falcon 9 rocket, reducing its price by about 30%.

5.   Small Satellite Launch Technology

Due to the rapid development of small satellite technology, military and commercial aerospace have developed a strong need to launch small satellites. As a result, commercial aerospace companies are being strongly encouraged to put forward innovative programs to compete for this market.

5.1 XS-1 and Small Satellite Launching Technology

In 2014, the Defense Advanced Research Projects Agency (DARPA) launched an XS-1 project to develop the Experimental Spaceplane. The goal of the XS-1 project is to develop a reusable booster vehicle to verify rapid response and low cost launch capabilities. It has two stages, the first of which can be reused. The upper stage is expendable. The project requires the first stage to fly 10 times within 10 days without the upper stage or payload. The final goal is for the entire system to deliver a payload greater than 1,362 kg into LEO with an inclination of 90° and an altitude of 185 km. The cost per launch is 5 million

US dollars. DARPA then announced three Phase 1 awards for initial studies of the XS-1 concepts. In addition to Boeing and its partner Blue Origin, DARPA gave awards to Masten Space Systems, (working with XCOR Aerospace) and Northrop Grumman (working with Virgin Galactic).

In April 2016, DARPA issued a call for proposals for Phase 2 and 3 of the program. Boeing, Masten, and Northrop Grumman all submitted proposals for Phase 2, but DARPA also allowed other companies to compete. It did not disclose the number of proposal it received. Phase 2 of the XS-1 program will cover the development of the vehicle and ground tests through 2019, with a series of 12–15 test flights for Phase 3 in 2010. DARPA announced on May 24, 2017 that it had selected Boeing to develop Phase 2 and 3 of the XS-1 program.

Boeing's proposal for the first stage, called Phantom Express, was developed by its Phantom Works. This booster vehicle will take off vertically, with the upper stage carrying a satellite payload mounted on top of the fuselage. After releasing the upper stage, the suborbital vehicle will glide back to a runway landing. In Phase 3, the Phantom Express will have to reach a speed of up to Mach 5. The following test flights will go up to Mach 10, and at least one test flight will carry an upper stage that would place a demonstration payload into orbit.

In Phase 1 of the XS-1 project, Boeing partnered with Blue Origin, with the expectation that the latter would provide an engine for the spacecraft. Phantom Express has decided to use the AR-22 engine developed by Aerojet Rocketdyne on the basis of the Space Shuttle Main Engine (SSME). In a statement, Aerojet Rocketdyne announced that it was providing two such engines 'with legacy shuttle flight experience' using parts from both the company's and NASA's inventories for earlier versions of the SSME. Phantom Works claimed that the Aerojet Rocketdyne engine was chosen as it offers flight-proven, reusable engines to meet DARPA's mission requirements.

DARPA chose Boeing's solution because the company has rich experience in developing space shuttles. The X-37B – a small unmanned space shuttle developed by Boeing's Phantom Factory – has been in orbit four times.

Secondly, in March 2014, DARPA selected Boeing to develop a carrier rocket launched from an F-15 aircraft for its Airborne Launch Assist Space Access (ALASA) program, aiming to place satellites weighing up to 45 kilograms into orbit for 1 million US dollars per launch, on 24 hours' notice. However, ALASA suffered problems related to its use of an unconventional 'mixed monopropellant' called NA7 – a mixture of nitrous oxide and acetylene. Ground tests found that NA7 was less stable than expected, so in November 2015, DARPA changed the goals of ALASA to continue testing NA7, scrapping the development of the launch vehicle. In its announcement of the XS-1 award,

DARPA stated that autonomous flight termination systems and related autonomous flight technologies developed as part of the ALASA program will be applied to Boeing's Phantom Express vehicle.

Finally, Boeing has significant economic strength. A DARPA spokesman stated that the value of the award to Boeing is 146 million US dollars. This is obviously not enough for the completion of the project. The award is structured as a public-private partnership, with Boeing also contributing to the overall cost of the program. However, Boeing has declined to disclose its contribution involving market competition.

As for the relationship between the XS-1 project and the US reusable launch vehicle, the development of US reusable space vehicles has been beset with difficulties. For example, the shuttle orbiter is too expensive to be viable. Under the pressure of cutting operation and supply costs for space stations, NASA has explored the National Aerospaceplane (NASP) with horizontal-takeoff and horizontal-landing (HTHL), X-33 and VentureStar with vertical take-off and horizontal landing (VTHL), and DC-X with vertical take-off and vertical landing (VTVL), but none of these projects achieved success. As the main mission shifted to exploring the Moon, asteroids, and Mars, NASA lost interest in developing reusable space vehicles.

Since it is responsible for a lot of military satellite launches, the US Air Force has had to keep an eye on reusable launch vehicles, but its fate remains poor. The USAF initially proposed the concept of a Hybrid Launch Vehicle (HLV), and awarded research contracts for ARES rockets to Northrop Grumman, Lockheed Martin and Andrews Aerospace in 2006. However, the Defense Authorization Act of Fiscal Year 2007 (NDAA 2007) terminated the ARES program. Keeping to its original intention, the USAF planned a series of programs for the demonstration and verification of the Reusable Booster System (RBS) in 2010. As the core technology behind some reusable launch vehicles, the RBS is consistent with the HLV. Pathfinder –the first RBS for demonstration and verification – was slated for testing in 2013. The RBS series aimed to halve launch costs on the premise that there were at least eight launches per year. The configuration of RBS included a reusable first stage and an expendable upper stage. Two RBS models were to be developed – one equipped with a reusable first stage and a cryogenic upper stage for medium-sized launch missions, and the other equipped with two reusable boosters, a cryogenic core stage, and an upper stage for heavy-duty launch missions. In 2011, USAF awarded contracts to the three companies again, but the program was short-lived. By October 2012, the Pathfinder was canceled due to a limited Air Force budget.

It is not difficult to see how closely DARPA's XS-1 program is related to the Pathfinder program. Jess Sponable, the XS-1 program manager, was chief engineer of the Pathfinder program at the Air Force Research Laboratory (AFRL) before joining DARPA in

December 2012. At the time when the Air Force's budget was being set, it seemed wise to transfer the task of developing RBS to DARPA, which has a certain amount of money and a degree of flexibility in management.

For now, documents released by DARPA about the XS-1 program make it clear that its primary mission was to launch satellites, and had looked into the possibility of putting them into commercial operation. Of course, it may evolve into a weapon platform in the future. Due to the limitation of the current delivery capacity, its combat ability is insufficient for it to be used as a weapon.

Although DARPA awarded Boeing the XS-1 contracts for phases 2 and 3, Virgin Galactic announced in February 2015 that it had rented a new plant in Long Beach, California, for the design and construction of its small satellite launch vehicle – the LauncherOne. This is a two-stage rocket designed to launch small payloads under 227 kg at a cost of under 10 million US dollars. Like Virgin Galactic's manned spacecraft, it will be launched by its own White Knight 2. Virgin Galactic, invested by Virgin Group and Qualcomm, has signed contracts with several companies to launch satellites including satellite-based Internet service provider OneWeb.

XCOR Aerospace was involved in the XS-1 program. Based at the Mojave Air and Space Port, California, it was a start-up company that had been engaged in the development of rocket engines and spacecraft since 1999. Like Virgin Galactic, it planned to launch a new suborbital commercial space travel program. For more than a decade, XCOR had been developing reusable rocket engine technology for commercial use. It built several generations of rocket engines, launched several rockets on the ground, and completed a series of flight tests on two spacecrafts driven by its rocket engines. XCOR's Lynx rocketplanes were powered by four liquid rocket engines with multiple ignition ability, and were made of lightweight materials. Lynx Mark I spacecraft could fly at an altitude of 61 km, at a cost of 95,000 US dollars to passenger. Lynx Mark II could take visitors to an altitude of 103 km and stay in space for five to six minutes at a cost of 220,000 US dollars. More than 200 bookings were made worldwide.

At the same time, XCOR and NASA were collaborating on a state-backed research and development program to build commercial aerospace manufacturing facilities and a Shuttle Landing Facility (SLF) at the Kennedy Space Center in Florida. Unfortunately, due to adverse financial conditions XCOR had to lay off most of its employees, and finally ceased operational entirely.

The Stratolaunch System – a commercial aerospace company founded by Microsoft co-founder Paul Allen – is building the world's largest aircraft with a length of 72 meters and a wingspan of 117 meters (longer than the standard football field), and a tail height of 15 meters. It is powered by six Boeing 747 jet engines of the same type, and a total of

28 wheels. This aircraft weighs 227 tons, and can carry 113 tons of fuel. Its maximum take-off weight has reached an astonishing 590 tons. However, the purpose of this huge aircraft is not to carry passengers, but to launch rockets. It will carry a rockets to an altitude of 9,100 meters and then launch satellites into low-Earth orbit.

Stratolaunch partnered with Orbital ATK to launch a Pegasus XL rocket and carry three rockets to launch at one flight. In July 2017, the plane was officially unveiled in California. After a series of tests, its maiden flight was made in 2019.

5.2 Zero2Infinity's Rocket Balloon

Based in Barcelona, Zero2Infinity is a Spanish company that develops near-space balloons for research and engineering customers. Having initially worked on testing high-altitude balloons to provide near-space tourism, the company is now focusing on developing Bloostar – a small satellite launch vehicle that uses high-altitude balloons as the first stage of a satellite launch, after which the rocket will put the satellite into orbit.

Astrophysicist James A. Van Allen et al. introduced the concept of the 'rocket balloon' in 1949. Its purpose is to send a rocket into space while allowing it to consume no extra fuel in the lower atmosphere. Although the technology was not used until the late 1950s, it did not develop because it could not be controlled in the lifting phase of the balloon, and therefore it could not be predicted where the rocket would fall after it was separated from the satellite.

Bloostar also adopted the concept of the rocket balloon. A helium balloon carries the small satellite and the rocket needed in the next phase to an altitude of around 30 kilometers. At this altitude, the air drag is already very small. Once in place, the rocket will be released from the balloon, and its liquid fuel rocket engine will be ignited to bring the small satellite into orbit.

Compared with a ground launch rocket, the biggest advantage of launching a rocket at a high altitude in thin air is that it can reduce the increment of speed required for entering orbit, due to the lower air resistance and gravity loss, and the fact that the engine nozzle can work under optimal conditions. In addition, the thermal environment of a high-altitude launch rocket will also be greatly improved, reducing the requirements for its thermal protection system. At the same time, after the launch of the rocket, the high-altitude balloon can also be used as a communication relay station, thus reducing dependence on the ground monitoring and control communication station.

The Bloostar rocket uses a parallel three-stage configuration. The first and second stage adopt a concentric toroid structure. The outermost first stage is a toroid wrapping on the second stage (a smaller toroid), and the center is the third stage with the payload fixed on it. Bloostar uses LOX/methane engines. The first stage includes six engines, each producing 15 kN of thrust. The second and third stages use the same engines, each producing 2 kN of thrust. The second stage includes six engines, and the third stage includes one engine. All engines are propellant pressure-fed, built partially by 3D printing technology with a total weight of only 4.9 tons.

On March 1, 2017, Zero2Ifinity lifted the small Bloostar rocket up to an altitude of 25 kilometers near the Spanish coast by balloon. After successfully realizing separation and ignition, recovery was completed at sea. However, controversies abound online as to whether the test was successful.

Zero2Infinit accepts launch bookings online. It has received an order intention worth 250 million euros, but has not disclosed the launch price. Of course, in the fierce market competition for small satellite launches, whether Zero2Ifinity's technological innovation scheme has advantages in terms of price and reliability remains to seen.

5.3 Rocket Lab's Electron

Rocket Lab is an American start-up focused on launching small, low-cost rockets. Its development can be a useful lesson for private rocket companies in China.

The company was founded in 2006 by New Zealander Peter Beck, and was the first to launch a rocket in the southern hemisphere, in 2009. In 2013, the Rutherford liquid engine was tested, and the Electron rocket project was proposed in the same year. In 2015, the Rutherford engine was fabricated largely by 3D printing, and a launch site was selected in New Zealand. In 2016, the engine was certified and the second stage of the rocket was finalized. Later that year, the rocket launch site was completed.

On May 25, 2017, the Electron rocket was successfully launched for the first time. So far, several other launches have been carried out, all of which have been successful. The Electron rocket can send a 150 kg payload into sun-synchronous orbit.

The company's goals are flexibility, low cost (under 10 million US dollars per launch), and the capacity for ridesharing.

The first stage of the Electron rocket is connected in parallel with nine engines, each with a thrust of less than two tons. Low-thrust engines are easy to manufacture and can be 3D printed. They can also be made reliably with existing materials so as to save development time and cost. The motor is directly driven by a lithium battery for fuel injection. This saves on machinery and reduces the weight of the engine. Carbon fiber is used as rocket shell material to reduce the structural weight. Liquid oxygen and kerosene are used as rocket fuel, as they are easy to store.

Although there are many innovations in the Electron rocket, Rocket Lab's pioneering commercial space launch service mode is also worthy of attention. Rocket Lab released its online booking system for satellite launches (http://https://www. rocketlabusa.com/book-my-launch/) on its official website in August 2015, becoming the first commercial aerospace company in the world to offer such a service. The United Launch Alliance (ULA) released the Atlas rocket online evaluation system on November 30, 2016. The launch mission of the Electron rocket has been scheduled for 2020. According to the information on the website, the rocket can send 24 3U CubeSats satellites and eight 1U CubeSats satellites into lower-Earth Orbit (LEO) per launch. Clients can log in to the company's website and access the online booking system to view the launch time of the rocket and its scheduled orbit (it can currently reach LEO with an inclination of 45°, and the circular sun-synchronous orbit with an altitude of 500 km), and select the weight and location of the payload to be launched. After an online booking, Rocket Lab will call the subscriber to confirm the order details. The launch price of CubeSats satellites depends on the number of remaining payload positions and the time of launching the rocket. The more the remaining payload positions or the longer the time of the launch, the cheaper the price. The launch price of 1U CubeSats is 50,000–80,000 US dollars, and that of 3U CubeSats is 20,0000–25,0000 US dollars. In this way, the online booking system has disrupted the traditional satellite launch ordering procedure.

After establishing a foothold in the small launch vehicle market, Rocket Lab is now planning to enter the small satellite field with a Photon platform that the company says can get customers into orbit faster. The Photon platform allows customers to integrate a variety of payloads from Earth observation cameras and communications equipment, and get them into orbit in less time than if companies built their own satellites. The company sees Photon as being particularly well-suited to technological demonstration missions, where customers bring a payload they want to get into space quickly before using it in a larger constellation. Rocket Lab could also provide the payloads in addition to the platform.

Each Photon can carry up to 170 kg of payloads. The first Photon will launch no earlier than 2020, because the company's 2019 launch manifest is full.

References

[1] Huang Zhicheng. *Sky and Sky Vision* [M]. Beijing: Electronic Industry Press, 2015.

[2] Huang Zhicheng. *The Fourth Wave of Aerospace Science, Technology, and Society* [M]. Guangzhou: Guangdong Education Press, 2007.

[3] Gao Zhaohui, Zhang Puzhuo, Liu Yu, et al. 'Technical analysis of the vertical return reuse carrier rocket' [J]. *Journal of Astronautics*, 2016, 37 (2): 145–152.

[4] Wang Fang, Cheng Hongwei, Bloomberg. 'Analysis and Explanation of the Successful Recovery of the 'Falcon 9' Launch Vehicle's Offshore Platform' [J]. *Journal of the Academy of Equipment*, 2016, 27 (6): 69–74.

[5] Squirrel. 'Multiplexing Technology and Prospect of 'Falcon 9" [J]. *Space Exploration*, 2016 (2): 25–29.

[6] Xi Shui. 'Lessons from Falcon 9' [J]. *Space Exploration*, 2016 (2): 30–33.

[7] Long Xuedan, Yang Kai. 'Analysis of the highlights of the Falcon 9 rocket mission' [J]. *International Space*, 2017 (2): 76–79.

[8] Wang Xin. 'The 'Raptor' rocket engine: the core of SpaceX's Mars colonization plan' [J]. *Space Exploration*, 2016 (11): 29–34.

[9] Hu Dongsheng, Zheng Jie, Wu Shengbao. 'Analysis of the New Glenn rocket compared with the Falcon Heavy rocket' [J]. *International Space*, 2017 (6): 45–48.

[10] Zhang Xuesong. 'The New Shepard Rocket: A Historic Vertical Soft Landing' [J]. *Space Exploration*, 2016 (2): 21–24.

[11] Zhang Xuesong. 'The 'Blue Origin' Space Hall of Fame' [J]. *Space Exploration*, 2016 (11): 35–39.

[12] Mu Yu, Wei Wei. 'Analysis of the Vulcan Rocket's technical scheme and low-cost control measures' [J]. *China Aerospace*, 2016 (7): 10–15.

[13] Fan Ruixiang, Ma Zhonghui, Yi Yi, et al. 'The Ariane 6 Carrier Rocket Plan and its Meaning for the Development of China's Carrier Rockets' [J]. *Missiles and Space Vehicles*, 2014 (6): 27–30.

[14] Federal Aviation Administration. *The Annual Compendium of Commercial Space Transportation: 2017* [R/OL]. https://brycetech.com/downloads/FAA_Annual_Compendium_2017.pdf.2017.

[15] BE-4 Rocket Engine [ER/OL]. https://www.ulalaunch.com/uploads/docs/BE-4_Fact_Sheet_Web_Final_2.pdf.2016.

CHAPTER 8

EARTH-TO-ORBIT TRANSPORTATION SYSTEMS

Earth-to-orbit transportation systems are extremely important for human space exploration, and their technology is very complex. This means that they have become the most challenging and interesting high-end field for aerospace start-ups. For this reason, space start-ups have a wide range of innovation possibilities.

1.   The History of Earth-to-Orbit Transportation Systems

An Earth-to-orbit transportation system is a means of transportation between the ground and space for conveying people and goods. At present, it refers to the transportation between the ground and the low-Earth orbit (LEO), the ground and the Moon orbit, the ground and the Moon's surface, and the ground and Mars. In the future, it will expand between the ground and other planets. Because spacecraft have to re-enter the atmosphere, they have to withstand severe heat in order to return to the ground in safety.

According to the mission of the system, its requirements are as follows:

1) The ability to complete tasks

   The number of crew and the weight and size of cargo to be loaded must meet the mission requirements, and must be delivered to the destination as well as safely returned. The comfort level of occupants, the rendezvous and docking ability, and the life-saving and emergency return ability of the crew must also meet the mission requirements.

2) High safety and reliability

   In the history of manned space flight, fatal accidents have caused huge economic losses and delayed the schedule of the program. Ensuring safety and reliability is critical for the Earth-to-orbit Transportation System. In addition, improving safety and reliability is subject to funding and schedule constraints. Therefore, in the process of developing an Earth-to-orbit transportation system, it is necessary to deal properly with the relationship between improving safety and reliability and realizing technical progress, as well as between the cost and the schedule.

3) Low transportation costs

For an Earth-to-orbit transportation system, the cost per flight is required to be low for manned launch. The cost per kilogram of cargo is required to be reduced for payload launch. The total cost of the fixed transportation weight should be reduced for expendable launch, while the total cost of the entire life required should be reduced for recoverable launch. The total cost includes development, manufacturing, and maintenance.

4) Good performance

This includes operation and maintenance, short ground turnaround time, less auxiliary equipment, and facilities for launch and return.

From a technical point of view, there are now three types of Earth-to-orbit transportation system: the vertical take-off and vertical landing (VTVL) spacecraft, vertical take-off and horizontal landing (VTHL) spacecraft, and horizontal take-off and horizontal landing (HTHL) spaceplane. The first two configurations have been applied, and each has its own advantages. As for the spaceplane, it is still in the pre-research stage.

Since Earth-to-orbit transportation systems are integral to the exploration of space, and the technology is very complicated, governments of aerospace countries have invested huge manpower and funds to develop systems that can meet the above requirements. Although many achievements have been made, there are also many lessons to be learned. In recent years, the rise of commercial aerospace has led to many new programs being developed, providing a new growing space for Earth-to-orbit transportation systems.

The development of Earth-to-orbit transportation systems has gone through three phases.

The first phase uses expendable spacecraft to challenge the basic technologies of manned space flight. These basic technologies are the first step towards space travel for human beings, and are also the foundation for building a large system for space station engineering. In the context of the space race in the Cold War, both the United States and the Soviet Union adopted expendable spacecraft based on ballistic rocket technology to send people into space in order to save time. This led to a high-volume utilization efficiency of the spacecraft, a smaller requirement for the rocket's carrying capacity, a lower technical risk and development cost, and easily guaranteed progress.

As technologies have become more mature, spaceships are improving in the process of development. At the same time, spacecraft are particularly suitable for landing on the Moon, as it has no atmosphere. Therefore, manned spacecraft and cargo spacecraft have become the main transport vehicles between the Earth and the Space Station, and are also the only type used to land on the Moon.

During the construction of spacecraft, it was found that the current technology had some shortcomings for the transportation from Earth to low-Earth orbit (LEO):

1) Large overload

   As the lift-to-drag (L/D) ratio increases, the maximum overload of the return vehicle also decreases. Early spaceships, such as the Soviet Union's Vostok, were spherical in shape and performed a purely ballistic re-entry. The maximum overload of re-entry was 8-9gG. With further development, the L/D ratio of the spacecraft increased to 0.15–0.30, with a maximum overload during re-entry of around 4G. If the overload is too large, it can affect the comfort and health of occupants.

2) Poor lateral maneuverability

   The lateral maneuvering distance increases with the increase of the L/D ratio. Generally, a spacecraft can only move 200–300 kilometers in a horizontal direction. When orbit maneuverability is given, the spacecraft has to wait a long time to return from orbit to the predetermined landing site, which makes emergency rescue more difficult. If the L/D ratio could be increased to widen the lateral maneuvering distance, the waiting time in orbit can be shortened. At the same time, poor lateral maneuverability also poses great difficulties for emergency rescue in the ascending section.

3) Wide dispersion of the landing site

   In the early days, the dispersion of the spacecraft during normal landing could reach tens of kilometers. Later, with lift control, the dispersion of the landing site was shrunk. In cases of emergency return, the landing site of the spacecraft is more difficult to control, especially when the guidance, navigation, and control system of the spacecraft fails. To make the spacecraft rotate and conduct pure ballistic re-entry, the landing site can deviate from the predetermined landing site by about 1,000 kilometers.

4) With the expansion of the scale of manned space activities, the total cost of an expendable spacecraft with a small payload capacity is larger. Although purchasing multi-launches in batches can reduce costs significantly, space launch costs are still relatively high.

The aim of the second phase is to continue to develop and use spacecraft and partially reusable spaceplanes to prepare for large-scale exploitation and exploration of space. The space shuttle is a combination of space technology and aviation technology. The successive development of American and Soviet space shuttles represents a major breakthrough of the technologies for Earth-to-orbit transportation systems. The space shuttle adopts a reusable orbiter with delta wings, which makes the overload of re-entry less than 3G, and also increases the lateral maneuvering distance to 1,500–2,000 km. It is capable of landing horizontally like a plane. The payload capacity of the space shuttle is large, and the orbiter can be reused. As the scale of manned space activities is increasing, the space shuttle can provide large payload capacity and volume, which is convenient for extensive space science and technology experiments. Easy-to-install robotic arms complete the tasks of satellite recovery and maintenance.

There are also many serious problems in the use of the space shuttles, such as failure to achieve the purpose of reducing intended transportation costs; inconvenient use and maintenance; long ground turnover time; low launch frequency; and low volume utilization efficiency, which will be discussed in depth later.

The aim of the third phase is to meet the future development and utilization of space resources. All major aerospace countries are working on advanced reusable Earth-to-orbit transportation systems, and striving to improve safety and reliability. They are also increasing mission adaptability, reducing transportation costs, and improving use performance. In this phase, in addition to the continuing development of spacecraft and space shuttles, the progress of aerospaceplanes will also be put on the agenda.

The aerospaceplane simplifies the launching operation due to horizontal take-off and horizontal landing (HTHL), reduces the manpower required for launching, and cuts the ground turnaround time, thus bringing down transportation costs. The aerospaceplane adapts well to military missions. It can return to the take-off base after circumnavigating the Earth, and can approach a target from multiple directions. If it could take off from an ordinary airport, it would have very important military value. Due to the use of air-breathing engines in the atmosphere, the aerospaceplane can make full use of the oxygen in the atmosphere, and can also use the atmosphere to provide lift in the ascending phase. The air-breathing engine has a lower energy density, and has a longer life than a rocket engine. It is also easy to maintain. Airplanes derived from the aerospaceplane can be developed into hypersonic strategic reconnaissance aircraft, interceptors, and bombers in the military. They can also be developed into hypersonic transport aircraft and passenger aircraft for civil purposes.

Looking to the future, the above three types of transportation system (the spacecraft, space shuttle, and aerospaceplane) have their own potential, but they must each develop reusable technologies to reduce transportation costs.

2. Durable Spaceships

From the initial phase of manned space travel to the present, expendable spacecraft have been widely used.

2.1 An Introduction to Spacecraft Development

The Soviet Union has developed three generations of manned spacecraft. The Vostok (first generation) had a total weight of 4.7 tons, and sent an astronaut into space for the first time. The second generation, called Voskhod, was an improved version of the Vostok. The Soyuz series (third generation) included five models – the Soyuz, Soyuz T, Soyuz TMA, Soyuz TMAA, and Soyuz MS, and successfully completed a large number of tasks. The Soyuz TMAA was an improved version of Soyuz TMA, as required by the International Space Station (ISS). The total weight of the Soyuz TMAA increased from 7,070 kg to 7,250 kg, and the number of flight days (including the time of docking on the ISS) increased from 180 days to 200 days. The Soyuz MS series is the latest upgraded version of the Soyuz spacecraft with improved communication and navigation system. The Soyuz spacecraft is made up of three modules: the orbital, the re-entry, and the service module. The Soviet Union also developed the Progress cargo spacecraft.

The early US manned spacecraft also came in three generations. The first generation was the Mercury spacecraft, weighing only 1.35 tons. The second generation – the Gemini series – was developed for the Apollo mission to the Moon. It was made up of a re-entry and an adaptation module, with a total weight of 3.7 tons. The third generation was the Apollo series. The United States launched a total of 17 Apollo spacecraft, including 11 manned flights: one Earth orbit, three lunar orbits, and seven Moon landing flights. The Apollo spacecraft consisted of a command and service module (CSM) and a lunar module (LM). It weighed 44.7 tons and could carry three astronauts. During the Moon landing, two astronauts drove the lunar module to the Moon, and one astronaut kept the CSM in lunar orbit. After the Apollo program, the United States completed the Skylab program using the CSM of the Apollo spacecraft. In July 1975, the Apollo CSM and the Soyuz capsule realized rendezvous and docking.

As a part of the Constellation Program (CxP) proposed by Bush administration for American astronauts to return to the Moon, NASA and Lockheed Martin are now developing a new-generation manned Orion spacecraft. Although Barack Obama ended the CxP when he took office, some of its key programs remain, including the Orion spacecraft, which may be used to land on Mars and asteroids. Many features of the Orion spacecraft and the shape of the crew module are similar to those of the Apollo, but its pressurized capsule has a volume of 20 cubic meters, a bottom width of about five meters, and a weight of about 23 tons. It can send four astronauts to Mars.

As well as the above-mentioned manned spacecraft invested by the US government, two private companies are investing in their own versions. One is SpaceX's Dragon 2 spacecraft. SpaceX's first successful cargo ship, the Dragon, was put into use and successfully reused. SpaceX is now building a manned spacecraft – the Dragon 2 – based on it. The other is Boeing's CST-100 manned spacecraft. Both spacecraft plan to send astronauts to the ISS in 2018.

In addition to SpaceX's Dragon 2 spacecraft, the Cygnus cargo spacecraft developed by Orbital ATK has also been put into cargo service to the ISS. Japan has developed and operated the HTV cargo spacecraft, and Europe has developed and operated the ATV cargo spacecraft with the payload capacity of up to seven tons. Meanwhile, China has developed and operated the Shenzhou manned spacecraft and Tianzhou cargo spacecraft.

2.2 The Progress of Commercial Spacecraft

The life of the ISS may be extended to 2024. Due to a break of several years from the retirement of the space shuttle to the desertion of the ISS, the United States decided to develop a new-generation Earth-to-orbit transportation system in the private sector to take over the LEO transport service. This will be the largest transformation in the history of American manned space flight. Private companies will become the main force in R&D and manufacturing, while NASA will play the role of 'Party A'. The management mechanism of private space companies is flexible and efficient, and it will be relatively easy to achieve a short development cycle and low cost.

In fact, in order to solve the problem of transportation to the ISS after the retirement of the space shuttle, NASA developed the Commercial Crew and Cargo Program (C3P) in 2005, and launched the Commercial Orbital Transportation Services (COTS) in early 2006. At the beginning of 2010, the Commercial Crew Development (CCDev) program was launched to encourage private companies to develop reliable, cost-effective commercial space transportation systems that would replace the space shuttle's mission of transporting cargo and personnel to the ISS.

Commercial companies in the United States have been responsible for a number of past successes in the field of manned space travel. NASA itself has never been involved in building rockets or spacecraft, leaving the actual design and construction to private contractors such as Boeing and Lockheed Martin. The difference between NASA's current COTS project and the previous one is the way that funds are used. In the past, NASA has offered to pay for development in order to ensure profit, which is almost guaranteed, regardless of how the company performs. COTS pays a fixed amount of funds to these companies after bidding. If the development costs exceed the expenditure, the bidding company will bear the responsibility. Therefore, private companies bear the main risks.

NASA's COTS program aims to develop commercial supply services for the ISS. NASA finally signed agreements with two innovative private companies to deliver payloads to the ISS. One is SpaceX, which uses its Falcon 9 rocket and Dragon spacecraft to deliver cargo to the ISS. The other is Orbital ATK, which developed the Antares rocket and the Cygnus spacecraft to deliver cargo to the ISS.

SpaceX's cargo ship Dragon was launched on May 22, 2012. On May 25, 2012, it docked with the ISS and delivered its first 455 kg cargo. In March 2017, the Dragon completed its 10th mission to deliver supplies to the ISS, and returned to Earth after four weeks in space. On June 3, 2017, SpaceX launched its first reusable Dragon spaceship with the Falcon 9 rocket.

The Dragon is the only cargo spaceship that is capable of returning significant amounts of cargo to Earth. Equipped with a thermal protection shield, the spacecraft can withstand extremely high temperatures, and can land safely during the re-entry period. By comparison, other expendable spaceships burn up on re-entry into Earth's atmosphere, as they do not have a thermal protection shield. The Dragon spacecraft is 5.9 meters long, with a maximum diameter of 3.6 meters and a weight of only 4.2 tons. Its maximum launch payload mass is six tons, and the maximum return payload mass is three tons.

The design of Orbital ATK's Cygnus spacecraft inherits the proven spacecraft technology of Orbital Sciences and its partners, which is composed of a Pressurized Cargo Module (PCM) and a Service Module (SM). The purposes of adopting mature technology are to reduce costs, lower risks, and shorten the development cycle. The Cygnus spacecraft can deliver up to 2,700 kg of pressurized cargo to the ISS at one time, and its Unpressurized Cargo Module (UCM) can also deliver unpressurized cargo. The first demo flight of the Cygnus was conducted on September 18, 2013. In January and July 2014, it was launched to the ISS twice by the Antares 120 rocket for Commercial Resupply Service (CRS). The first night launch took place on October 28, 2014, but suffered a catastrophic anomaly resulting in an explosion shortly after take-off. On December 6, 2015, it was successfully launched, sending 3.5 tons of supplies and instruments to the ISS. On April 18, 2017, Orbital ATK launched the Cygnus spacecraft with the Atlas V rocket, and carried out the seventh cargo mission to the ISS.

In September 2014, NASA awarded contracts to Boeing and SpaceX to develop the Commercial Crew Transport Capability (CCtCap) for a joint manned mission to the ISS.

The number 100 in the name of the Boeing CST-100 spacecraft represents 100 kilometers, or the distance between the Earth and the low-Earth orbit (LEO), which means that it is designed for short-haul flights into space. The spaceship looks similar to the Apollo and Orion spacecrafts, but its volume is between their sizes, and it can carry seven astronauts at a time.

The innovation of the CST-100 includes the ship's weldless design, modern structure, and upgraded thermal control technology. The integrated design method reduces the overall weight of the spacecraft and speeds up the manufacturing of the module.

According to foreign media reports, due to a series of development problems, Boeing decided to postpone the first manned test flight of the CST-100 to August 2018. The first 'certified' flight of the CST-100 spacecraft to the ISS is expected in December 2018.

On May 29, 2014, SpaceX unveiled the latest Dragon V2 capsule in Hawthorne, California. It will launch into space atop a Falcon 9 rocket, and can carry a crew of seven astronauts into orbit. It is expected to launch manned flight by the end of 2018.

The shape of the Dragon V2 is obviously different from its cargo predecessor. It has a non-axisymmetric, relatively flat shape. Four groups of eight SuperDraco rocket engines (developed by SpaceX over a period of many years) are distributed in an X-shape on both sides of the cabin. This configuration is used to improve the lift-to-drag ratio at the time of re-entry, allowing the astronauts more comfort during re-entry, and reducing aerodynamic heating for reuse.

The landing solution used by the Dragon V2 crew capsule is a nail-biting landing method, rather like that of a helicopter. Using the SuperDraco engine described above, the space capsule can slow down or even hover at low altitudes, and then extend four landing legs to land anywhere. In order to increase safety, a parachute is also on board as a backup. These engines can also be used for emergency rescue during ascent. The Dragon 2 capsule will be reusable. No components will need to be replaced, including the heat shield outsole. It will simply need to refuel before being launched again. It is reported that the main structure of the capsule is expected to last for up to 10 flights before needing significant refurbishment. This gives the space capsule an edge in terms of economy. Dragon 2 is estimated to cost only about 20 million US dollars per flight – much less than NASA's purchase of a Soyuz seat.

On January 22, 2016, a video released by SpaceX shows the Dragon 2 space capsule successfully completing a hover test. A cable was attached atop of the model of the Dragon 2 capsule, and eight SuperDraco thrusters were fired at the same. The capsule hovered in the air for about five seconds using its own power. SpaceX intends to use the reverse thrust technology to slow descent for soft landings back on Earth, but will not use this technology when it initially carries crews. Instead, parachutes will be used to drop the Dragon 2 into the ocean.

SpaceX successfully launched the crewed Dragon 2 capsule from the Complex 40 launch pad of Cape Canaveral Air Force Station at 09:00 am on May 6, 2016. The main purpose of the launch was to test its latest crew abort system. During the launch, the capsule triggered an escape maneuver to simulate an emergency under intense aerodynamic stress. Therefore, the launch carried only one dummy. From takeoff to landing, the entire test took 99 seconds.

In July 2017, SpaceX announced a design change for the Dragon 2 capsule. Due to the safety risk involved in the landing legs passing through the capsule's heat shield, plans for a propulsive Dragon landing were abandoned. At the same time, judging by the example of the reusable Dragon cargo capsule launched on June 3, 2017 and returned on July 2, 2017, the cost of reusable spacecraft is no lower than that of expendable spacecraft due to the high maintenance cost. However, the cost is still expected to be reduced in the future. Efforts of to reuse the Dragon 2 capsule and reduce costs still need consistent practice and improvement.

3. Space Shuttles for Civil-Military Integration

The space shuttle is a space vehicle in the shape of an aircraft. It is powered by a rocket engine, and can land horizontally between the Earth's surface and space. Many modern scientific and technological achievements are concentrated in this vehicle, and it is the amalgamation of rocket, spacecraft, and aircraft technology.

3.1 The Successes and Failures of the Space Shuttle

In the 1970s, in order to compete with the Soviet manned space flight, the US Congress approved the Space Transportation System (STS) using the space shuttle program in early 1972. It took nine years and cost about 10 billion dollars. In April 1981, the United States finally launched the first space shuttle, Columbia, into space. Translated as 航天飞机 in Mainland China and 太空梭 in Taiwan, the space shuttle is just one type of spaceplane.

The space shuttle's components include an Orbiter Vehicle (OV), a pair of recoverable solid propellant boosters (SRBs), and an expendable external tank (ET). The United States built five of these shuttles. The first, Columbia, successfully completed its first orbital flight on April 12, 1981. It crashed on re-entry on February 1, 2003. The orbiter Challenger also blew up after lift-off on January 28, 1986. The remaining orbiters – Discovery, Atlantis, and Endeavour – retired in 2011.

The Soviet Union also developed a successful space shuttle called Buran. The shape of its orbiter was very similar to that of the US space shuttle, but its main engine was not fired on the ground. Instead, it was fired only after being sent to an altitude of 150 kilometers by the carrier rocket. The Buran made a successful unmanned orbital test flight in 1991, but never flew again due to a shortage of funds.

The space shuttle is made of three main components: an orbiter vehicle (OV) for crew members; a large External Tank (ET) that provides fuel for the main engine; and two Solid Rocket Boosters (SRBs) that provide most of the lift for its first two minutes. The other two components can be reused, but the ET will be burned up in the atmosphere after each launch.

As the core component of the space shuttle, the orbiter is 37.24 meters long and 17.27 meters high, with a 29.79-meter wingspan. Its front section is the crew compartment, and its middle section is the cargo bay. The upper part of the cargo bay can be opened like clamshell. Attached to the cargo bay is the Remote Manipulator System (RMS), also known as Canadarm – a remote-controlled robotic arm built in Canada, that is used to launch and retrieve payload such as satellites and probes. Upper stage rockets in the cargo bay can also be used to launch the spacecraft into higher orbits. Recovered satellites and other spacecraft can be repaired in the cargo bay. Its rear section has a vertical tail, three Space Shuttle Main Engines (SSMEs), and two Orbital Maneuvering Systems (OMS). Depending on the fuel from the ET, the SSMEs starts working at take-off, each generating a thrust of 1,668 kN. The double-delta wings are on the outer sides of the middle and rear sections of the orbiter. About 20,000 heat-resistant tiles are attached to the orbiter's head and the leading edge of the wing to protect the orbiter from being destroyed by high temperatures of 600–1,500°C generated by aerodynamic heating on re-entry. There are also 44 small engines in the nose cone and tail of the orbiter for slight orbit adjustment. The ET is 46.2 meters long and 8. 25 meters in diameter, and can hold more than 700 tons of liquid hydrogen and oxygen propellant. There are two Solid Rocket Boosters (SRBs), which are connected to both sides of the ET. They are 45 meters long and 3.6 meters in diameter. Each of them can generate 15, 682 kN of thrust.

The space shuttle was a major symbol of innovation for manned spacecraft in its attempt to become reusable. Its main advantages, in addition to being able to be reused multiple times, is its ability to carry seven people and nearly 30 tons of cargo into space. Not only that, the weight of payload from orbit to Earth is 14.5 tons on re-entry. These features play an irreplaceable role in the construction and installation of the ISS. In the past 30 years, the space shuttle has not only completed the construction of the ISS, but has also accomplished a lot of tasks such as deploying satellites, and launching space probes and the Hubble telescope. It has also conducted a series of successful projects including satellite recoveries and space repairs. It has transported more than 1,360 tons of payload and more than 600 astronauts into space.

Due to a lack of experience, as well as the limitation of development costs and schedules, the space shuttle also revealed serious defects after it was put into operation. First of all, there were many potential risks. The three main components of the space shuttle – the external tanks (ET), Solid Rocket Boosters (SRBs), and orbiter vehicle (OV) – were connected in parallel, and many hidden dangers were found after it was put into operation. While the space shuttle continued to set records, it was involved in some of the most tragic moments in human space exploration.

Seven astronauts lost their lives in the explosion of the Challenger in 1986 and the disintegration of the Columbia in 2003, meaning huge setbacks for American manned space flight. After careful investigation and analysis, it was finally confirmed that the Challenger accident was caused by the failure of the synthetic rubber O-ring seal at the joint of the SRB on the right side. The reason for the Columbia accident was that a foam cube on its ET surface fell off and hit its left leading edge, causing a wing to burn.

Secondly, in the absence of considering issues that might arise in operation, it failed to reduce transportation costs. As for its external heat-resistant tiles, damage was caused on each flight, and maintenance and replacement works proved to be laborious. As the energy density of a rocket engine is very high, it needs to be disassembled and cleaned after each flight, and many parts need to be replaced (such as the turbine pump), before it can be reused. This resulted in high launch and maintenance costs. It was reported that the space shuttle costs as much as 500 million US dollars per flight, and the cost of transporting a kilogram of payload to low-Earth orbit (LEO) was around 10,000 US dollars. It also required a lot of time-consuming and labor-intensive overhaul after each return, resulting in 51% of recurring costs being hardware costs (including expendables), 4% being propellant costs, and 45% being operation expenses.

Finally, the launch frequency was far from what was expected. Due to the long turnaround time on the ground, the launch frequency was low. The space shuttle was supposed to have a maximum lifespan of 20 years, and each shuttle should have been able to fly 100 times. In fact, the five shuttles flew only 133 times all together, and each flew no more than four times a year.

The most important finding was that it was more expensive to launch a satellite with a space shuttle than a rocket. As a result, NASA decided not to undertake the launch mission for commercial payload after 1988, and the total number of space shuttle flights per year was reduced to about nine. Although the number of tasks performed was nearly a quarter less than expected, damage and deterioration were still aggravated, resulting in a further increase in the cost of each maintenance.

Built as a collaboration between the United States, Russia, Europe, Japan, and Canada, the International Space Station launched its first Functional Cargo Block (FGB) in November 1998. Later, progress was very slow due to funding, technical issues, and program coordination. In particular, the crash of the space shuttle Columbia in 2003 greatly delayed the progress of the program. After the space shuttle began to be launched again in 2006, things went smoothly, and the ISS was finally completed in 2011.

After NASA had successfully developed the space shuttle, it spent a lot of money to develop the spaceplane and the single-stage LEO rocket plane, but all failed due to inadequate technical capacity. After the retirement of the space shuttle in 2011, the United States has to pay for seats on Russian manned spaceships to transport astronauts to the ISS.

3.2 The X-37B Small Military Aerospaceplane

Another form of the space shuttle is placed on top of a rocket, but this configuration is inevitably constrained by the current rocket carrying capacity.

The US military was the first to pay attention to the spaceplane. The Dyna-Soar (also known as the X-20) – developed by the US Air Force from October 24, 1957 to December 10, 1963 – was a spaceplane of this type. The program was subsequently canceled for a variety of reasons. In the 1980s, countries such as France and Japan began to research and develop this kind of spaceplane for shuttling. The French spaceplane Hermes was canceled, and the Japanese are continuing their research. The small unmanned spaceplane X-37b belonging to the US military has conducted four in-orbit tests. In recent years, the Dream Chaser spaceplane developed by Sierra Nevada Corporation (SNC) has been confirmed by NASA as one of the resupply systems for carrying cargo to the ISS.

The X-37B is based on the X-37 that was developed by NASA to test reusable space vehicles in orbit and re-entry environments. After former President George W. Bush proposed that the agency should concentrate on returning to the Moon, NASA decided to stop developing the X-37 program. In November 2006, this program was accepted by DARPA to co-fund the X-37B, of which the prime contractor is Boeing Integrated Defense Systems.

As a long-term strategic idea, military aerospaceplanes have been pursued by the US military since the 1950s. This is because the spaceplane has much greater longitudinal and lateral maneuverability during re-entry than the spacecraft, allowing it to land at multiple airports, and also to strike different targets over a wide area of the Earth. In addition, the spaceplane is easier to reuse because it has less overload during re-entry and landing.

The X-37B is an aerospace vehicle that tests military aerospaceplane technology. It is approximately one quarter of the size of space shuttle orbiter and weighs 4,989.5 kg.

It has a length of 8.92 m, a wingspan of 4.572 m, and a height of 2.926 m. The cargo bay measures 2.13 m × 1.22 m and its payload is about 227 kg.

The first difference between the X-37B and the space shuttle is that it is a UAV (Unmanned Aerial Vehicle). This means that no complicated life support system or life-saving system are required, which can reduce structural weight. Secondly, its power supply has been changed from the space shuttle's LH2/LOX fuel cell to a GaAs solar cell array and lithium-ion battery, which allows it to operate in orbit for a long time. Its fourth flight lasted for 718 days. Finally, it uses a storable and non-toxic aviation kerosene JP-8/hydrogen peroxide propulsion system (in its first flight it used a more mature nitrous oxide/hydrazine propulsion system), with a single thrust of 29.4 kN, making its orbit maneuverability much greater than that of the space shuttle.

Having learned lessons from the Columbia crash, the configuration of X-37B went back to the traditional way, with the orbiter vehicle on the top-stage of the rocket (like the X-20), avoiding the issue whereby the insulating foam on the external tank (ET) would break the orbiter's heat-resistant tiles. A fairing was designed specifically for the X-37B to reduce aerodynamic loads and improve vibration characteristics during the boost phase. Its interface can be used on a variety of rockets, including the Atlas V, Delta IV, and the new Falcon 9 from SpaceX.

Although the double-delta wings of the space shuttle are also used in the aerodynamic shape of the X-37B, the bluntness of the fuselage nose is larger. Instead of the single vertical stabilizer rudder of the space shuttle, it has to two angled tail fins. In this way, the yaw performance of the X-37B is improved, and the height of the whole spaceplane is reduced, so that it can still be put into fairing after the drag brakes are installed at the bottom of the fuselage. The cargo bay of the X-37B is equipped with two doors, which can be opened during in-orbit operation. The solar panels should be folded and placed in the cargo bay as limited volume. After entering orbit, the cargo bay will be opened automatically after the support arm is extended. Inside the door are radiant evaporators and solar panels. The radiant evaporator is designed to dissipate heat from the X-37B. Because it is directly exposed to the sun in space, its temperature rises rapidly. There is no air in space, so radiation is the only way to dissipate heat.

After completing its orbital mission, the X-37B will experience severe aerodynamic heating on re-entry. Obviously, it is necessary to install a Thermal Protection System (TPS) outside the fuselage, which uses second-generation American thermal protection materials. The head and leading edge are covered with Toughened Uni-piece Fibrous Reinforced Oxidation-Resistant Composite (TUFROC), developed by NASA's Ames Research Center, which has unique properties such as a light weight, low thermal conductivity, and better oxidation resistance. A large area of Toughened Uni-piece Fibrous Insulation (TUFI) material is used. In addition, as an experimental proposal, carbon/silicon carbide thermal protection material is partially used on the rudder surface.

The X-37B uses lighter and more advanced computers in its re-entry and return guidance, navigation, and control systems. Its navigation system uses a combination of inertial navigation and GPS, as well as the full Fly-by-Wire (FBW) system that has been widely used in modern aircraft, to enable autonomous re-entry, return, and landing. These improvements have greatly enhanced the vessel's reliability.

The X-37B flight test first focused on the technology of the X-37B itself, and then on the sensor technology that needed to be tested over a long period of time in orbit. These sensors for space reconnaissance and surveillance are highly confidential programs for the United States. Finally, on the fourth flight test, the Hall-effect Thruster (one of the advanced space propulsion systems) was tested. Some advanced materials were also taken on board to help NASA study their durability in space environments.

Regarding the future commercial application of X-37B, at a meeting of the American Institute Aeronautics and Astronautics (AIAA) in 2011, Arthur Grantz – Boeing's program manager for the X-37B – revealed a detailed design showing how a scaled-up version of the X-37B configuration is expected to be used to deliver cargo and crew to the ISS and other low-Earth orbit (LEO) destinations. The goal is to provide greater cargo capacity than the CST-100 spacecraft, as well as a possible, longer-term human transport capability. The future of the X-37B is divided into three steps. The first will use the current 8.8-meter-long aircraft for the demonstration flight to the ISS, and the X-37B will be launched in the 5-meter-diameter fairing of the Atlas V rocket. The next step will develop a larger vehicle, approximately 165 percent bigger than the current X-37B at about 14. 3 meters long, which will be sufficient to deliver larger Linear Replacement Units (LRUs) to the ISS, while reducing the risk associated with shuttle flights for astronauts. In the third step, a manned vehicle capable of carrying 5–7 passengers will be developed.

3.3 The commercial Dream Chaser spaceplane

On April 18, 2011, NASA announced the Commercial Crew Development Round 2 (CCDev-2) to help private US companies develop the Advance commercial crew space transportation system that would replace NASA's space shuttle. Boeing's award – the largest at 92. 3 million US dollars – was for their CST-100 capsule; Sierra Nevada Corporation (SNC) from Colorado received 80 million US dollars for the Dream Chaser spaceplane. SpaceX from California received 75 million for its Dragon spacecraft, while Blue Origin from Washington State received 22 million for its vertical take-off and virtual landing (VTVL) program. However, in 2016, NASA announced that Orbital ATK, SpaceX, and the Sierra Nevada (SNC) had won the second round of commercial cargo contracts for the ISS.

Because the design of the Dream Chaser spaceplane is in the HL-20 lifting body configuration developed by NASA for many years, it is a cooperative project with NASA itself. In 2004, SpaceDev announced that it would join the Vision for Space Exploration (VSE) proposed by NASA, and later the Commercial Orbital Transportation Services (COTS) project with the Dream Chaser. In 2006, the company signed a contract with NASA and officially decided to use HL-20 to develop the Dream Chaser. Although it failed to obtain the contract of COTS, it decided to continue cooperating with NASA to develop the project until SNC acquired SpaceDev in 2008.

'Lifting body' refers to an aircraft shape that does not rely on the traditional wing but on a fuselage-wing fusion body to produce lift. In the 1940s, aerodynamic scientists found that the wing and fuselage would produce aerodynamic interference, and also had the idea of using fuselage to generate lift. In the 1950s, when two doctors from the Ames Research Center of NACA (later NASA) were designing a warhead that could generate lift for autonomous control, they found that if the surface of the blunt cone (which was originally axisymmetric) was flattened a little, lift could be produced, and the re-entry orbit of the warhead could be controlled using this lift. In this way, the 'lifting body' concept was put forward. From the 1960s to the 1970s, NASA carried out a series of research projects and experiments on lifting bodies, such as MA2-F2, MA2-F3, HL-10, X-24A, and X-24B.

Shortly after the lifting body concept was proposed in the United States, the Soviet Union also started its own research. In response to the US development of the X-20 hypersonic power glider project, the Soviet Union began to implement a spaceplane project called EPOS (Experimental Passenger Orbital Aircraft), of which the MIG-105 was the demonstration plane. Later, in 1976, the first flight of the MIG-105-11 subsonic model was made, and the flight tests (totaling eight in all) continued sporadically until 1978. At this time, the orbital vehicle was decommissioned by the Soviet Union when the decision was made to proceed with the Buran Project instead. Later, the Soviet Union used the previous results to develop the reduction scale re-entry test vehicle of the BOR unmanned orbit rocket aircraft. This was similar to the X-23 PRIME (Precision Recovery Including Maneuvering Entry) re-entry vehicle in the United States and the ASSET (Aerothermodynamic Elastic Structural Systems Environmental Tests) re-entry vehicle in the 1960s. Between 1982 and 1984, BOR-4 made several test flights, and acquired the first BOR-4 photos obtained by a Royal Australian Air Force P-3 Orion reconnaissance aircraft on June 3, 1982, when Soviet ships were recovering a BOR-4 (No. Cosmos 1374) that had completed suborbital tests near the Cocos Islands.

NASA finally obtained information about the Soviet BOR-4 lifting body (including its shape, weight, and center of gravity) from photos taken by the Australians. At the end of 1986, a wind tunnel test for the BOR-4 model was conducted at NASA's Langley Research Center (LaRC). The results showed that it had good aerodynamic characteristics across its flight speed range from low subsonic speed to hypersonic speed. In 1986, NASA urgently needed a Crew Emergency Rescue Vehicle (CERV) for the proposed International Space Station (ISS) due to the fatal crash of the space shuttle Challenger. In the 1990s, NASA's LaRC was preparing to develop a Personnel Launch System (PLS), or Assured Crew Return Vehicle (ACRV), to improve flight safety at low operating costs. For this reason, NASA LaRC developed the HL-20 lifting body configuration.

Later, Johnson Space Center's X-38 program of superseded other programs to become the Crew Return Vehicle (CRV), sometimes referred to as the Assured Crew Return Vehicle (ACRV), of the International Space Station (ISS). The X-38 project was not implemented due to a lack of funding.

In addition to the HL-20 and X-38, NASA also adopted the lifting body configuration on the X-33. The X-33, developed by Lockheed Martin's renowned Skunk Works, was a half-scaled prototype of the Single-Stage-To-Orbit Reusable Launch Vehicle (SSTO RLVs) VentureStar. In March 2001, NASA canceled the 1.3 billion US dollar X-33 project because of insurmountable technical difficulties. NASA conducted numerous wind tunnel tests on all three projects, accumulating a large amount of aerodynamic data.

Like the space shuttle, the Dream Chaser also takes off vertically and lands horizontally (VTHL), but it is designed to launch from atop an Atlas V rocket. Since it does not need to carry a lot of cargo, the Dream Chaser is much lighter and smaller than the space shuttle. It is about nine meters long and seven meters in diameter, and weighs 11 tons. The structure of the Dream Chaser uses a large number of advanced composite materials. Its in-orbit power is provided by two thrust-adjustable solid-liquid mixed rocket engines, and its propellants are hydroxyl-terminated polybutadiene (HTPB, solid fuel) and nitrous oxide (oxidant). Both are non-toxic and easy to store. These engines were also used on Virgin Galactic's SpaceShipTwo. SNC has been exploring hybrid rockets for over a decade, conducting more than 300 ignition tests for hybrid rocket engines, and helping SpaceShipOne win the Ansari X prize.

On May 22, 2013, NASA Administrator Charlie Bolden met with a team at NASA's Dryden Flight Research Center as they prepared to test the Engineering Test Article (ETA). As an astronaut who had flown the space shuttle, Bolden also experienced the

Dream Chaser flight simulator. Since then, ETA has been tested through the ground, approaching, and landing stages. Unfortunately, an accident during its first test landing in 2013 prevented the team from signing a NASA contract for a commercial manned program in 2014. However, Sierra Nevada did not give up. It decided to develop the cargo Dream Chaser before the manned spaceplane. This Dream Chaser cargo version can be launched by a number of American launch vehicles including the Atlas V rocket. The wing of the lifting body can be folded and placed in a fairing with a five meter diameter. A cargo bay is added at the rear of the lifting body, which can transport 5,500 kg of cargo into orbit. As the Dream Chaser has the ability to rendezvous and dock with the ISS, it can quickly and conveniently deliver cargo and resupply, meeting NASA's requirements for payload to the ISS. It can also land at ordinary airports, be reused 15 times, and deliver critical cargo in a timely manner in emergencies.

4. The Aerospaceplane of the Future

The term spaceplane is an abbreviation of aerospaceplane. The spaceplane is a space vehicle that uses an air-breathing engine in the ascending stage to take off and land horizontally (HTHL) on a conventional runway. It has the appearance of an airplane, and travels between the Earth's surface and its orbit. Using oxygen in the atmosphere as an oxidant can reduce the weight and volume of the aerospaceplane and cut down operation costs. As an aerospaceplane can be completely reused like an ordinary plane, no equipment is needed to launch the rocket vertically.

4.1 The National Aerospaceplane Program (NASP)

On February 4, 1986, then President Ronald Reagan announced the implementation of the National Aerospaceplane Program (NASP), hoping to develop a spacecraft that could be completely reused, featuring single-stage-to-orbit (SSTO) and horizontal take-off and landing (HTHL). At the same time, Germany was working on a two-stage-to-orbit (TSTO) aerospaceplane. The main difference between the aerospaceplane and the space shuttle is that it uses an air-breathing engine in the major part of the ascending phase, so it can take off at an ordinary airport and perform maneuvers. In this way, it is suitable for military deployment.

From take-off to orbit, an aerospaceplane needs to accelerate from Mach 0 to about Mach 25. Generally speaking, the specific impulse of an air-breathing engine is much higher than that of a rocket engine from a very low flight Mach number to about Mach 15. Due to various limitations in structure and energy conversion efficiency, an air-breathing engine can only work within a range of flight Mach numbers. In order to get into orbit, it is necessary to combine a variety of air-breathing engines, often in combination with rocket engines.

Another option for the aerospaceplane is two-stage-to-orbit (TSTO). Qian Xuesen put forward the concept of using an air-breathing engine in the book *An Introduction to Astronavigation* published in 1962. He proposed using a large jet aircraft as the first stage vehicle and an aircraft equipped with a rocket engine as the second stage. He proposed 'taking off with the turbojet engine, speeding up to Mach 2. 0, and climbing to an altitude of 10 km; then starting the ramjet engine, and continuing to climb and accelerate until the limit is reached; then, the second-stage rocket will separated from the first-stage, and will start to lift off'. During the 1980s, many countries investigated this concept, including West Germany. In 1986, West Germany put forward a Saenger concept design for a TSTO spaceplane.

The advantage of SSTO is that it can operate like a normal aircraft, reducing operating and maintenance costs. The disadvantage is that the entire plane has to be subjected to severe aerodynamic heating during re-entry, which greatly increases the weight of the thermal protection system. At the same time, it is very difficult to accelerate the aerospaceplane from Mach 0 to about 25 using the SSTO propulsion system. Although the scheme of the TSTO increases the complexity of operation, its propulsion system can be configured separately for two stages to reduce difficulty. Only the upper stages re-enter the atmosphere after leaving orbit, reducing the weight of the thermal protection system.

As for the flight Mach number during two-stage separation, three schemes have been put forward: low, medium, and high. The low scheme chooses the high-subsonic flight Mach number separation in the existing civil aircraft, with a separation altitude of about 10 km. In the medium scheme, separation at Mach number 5–6 is selected, along with a separation altitude of between 30–40 km, as in the Saenger scheme above. The separation altitude of the high scheme is about 70–100 km at Mach 10.

Another advantage of the TSTO separation at Mach 5-6 is that its carrier technology will promote the development of hypersonic aircraft (including military aircraft and airliners).

In 2001, NASA and the US Department of Defense jointly put forward the National Aerospace Initiative (NAI), suggesting that the development of hypersonic vehicles in the United States should be divided into three steps: in the short term, researching supersonic/hypersonic cruise missiles aimed at crucial targets; in the medium term, focusing on the development of hypersonic bombers that can achieve global reach; and in the long term, building reusable vehicles that are affordable and capable of accessing space in good time.

At present, three kinds of hypersonic aircraft are being developed around the world. The first is the SR-72 hypersonic reconnaissance aircraft from Lockheed Martin. The SR-72 is different from the previous-generation SR-71 Blackbird reconnaissance aircraft. It is a UAV configuration, using a combination of a turbojet engine and a scramjet engine. After the turbojet engine accelerates the aircraft to Mach 3, the scramjet engine ignites and continues to accelerate to Mach 5~6. *According to Aviation Week*, Skunk Works said that the biggest difficulty encountered in the development of the SR-72 reconnaissance aircraft was how to coordinate the working conditions of two engines, that is, how to compensate for the speed gap between the turbojet engine and the dual-mode ramjet engine. Because the turbojet engine can only accelerate the aircraft to Mach 2.5, the minimum ignition speed of the ramjet engine is Mach 3, so Skunk Works is developing a ramjet engine that can ignite below Mach 3. At the same time, Skunk Works is working to improve the performance of the turbojet engine, so as to increase its working speed to Mach 4. It is also experimenting with the combination of a ramjet engine and a rocket engine.

The second hypersonic aircraft is the Skylon aerospaceplane that is being developed by the Reaction Engines Limited (REL) in the UK. It features two strong pre-cooling Synergistic Air-Breathing Rocket Engines (SABRE), and is designed to take off with a total weight of 325 tons and a payload of 15 tons. This means it can be used to deliver cargo, deploy or capture satellites, and rendezvous and dock with the ISS. Derived from the SABRE engine, the Scimitar engine will be used as the power for the civil hypersonic transport aircraft/airliner with Mach 5 in the LAPCAT-II program (Long Term Advanced Propulsion Concepts and Technologies Phase II).

In the middle of September, 2016, the US Air Force Research Laboratory (AFRL) revealed two models of the TSTO (two-stage-to-orbit) aerospaceplane concept based on the SABRE engine, one of which is first-stage reusable, with a total take-off weight of 159 tons, a take-off distance of more than 600 meters, and a payload of 2. 3 tons. Its first stage is powered by two SABRE engines with a propellant mass ratio of 0.43 (far below the first-stage reusable spacecraft powered by a rocket). When the speed reaches Mach 4.4 and the altitude reaches around 20 km, the vehicle is converted to rocket mode and climbs to altitude of 80 km. When the speed reaches Mach 8, the first stage is separated from the second stage. The second stage is pushed into orbit by the rocket engine, while the first stage returns to ground for reuse.

The third hypersonic aircraft is the Aerobic Vehicle for Hypersonic Aerospace Transportation (AVATAR) proposed by India's Defense Research and Development Organization (DRDO). It is a reusable, horizontal take-off, air-breathing SSTO aerospaceplane with a combined turbo-ramjet/scramjet/rocket cycle engine. It has a total weight of 25 tons (60% of which is liquid hydrogen fuel), and can send a payload of one ton into space once every launch. This aerospaceplane can perform 100 such missions during its life cycle, and can also be used as a hypersonic aircraft for air-to-ground attacks or reconnaissance, before returning to the base.

4.2 Scramjet & ramjet

In order to achieve hypersonic flight, multiple air-breathing engines should be combined, as well as air-breathing engines and rocket engines. Therefore, it is necessary to study a variety of air-breathing engines that operate in different speed ranges.

The working process of the turbojet engine is as follows: the air stream flows into the engine inlet at high speed; the multi-stage compressor decelerates and pressurizes the airflow, and then it is sent into the combustion chamber by the compressor and mixed with the injected fuel for combustion. The hot gas passes through the turbine to increase the air pressure, and then the high-pressure and high-temperature gas is sprayed out by the tail nozzle of the engine to generate thrust. At present, the general turbojet engine can only work at a Mach number slightly higher than 3. If the Mach number is higher, the turbine blades will be damage by the heat.

A ramjet is an air jet engine that draws the headwind into the engine and slows it down so that the air increases static pressure. Due to shock waves during supersonic flight, the velocity of the air in the inlet reduces to subsonic speed. At this time, the stagnation of the airflow can increase the air pressure more than ten times, which is much more than the increase of the pressure in the compressor of the turbojet engine. This means that the ramjet engine only needs the speed of ram air instead of the compressor. It is usually composed of three parts: intake, a combustion chamber, and a propulsion nozzle. As the ramjet has no compressor (no gas turbine is needed), it is also called an air jet without a compressor. When the ramjet works, the high-speed airflow blows directly towards the engine, expands, and decelerates in the engine intake with a rise in the air pressure and temperature, then enters the combustion chamber and combusts with the fuel, raising the temperature to 2,000–2,200°C or even higher. When the high-temperature gas passes through the narrow throat, it accelerates to the speed of sound again, then expands and rushes out of the inverted cone nozzle. It reaches supersonic speed, and finally discharges violently from the nozzle at a high speed to generate thrust.

In 1913, French engineer René Lorin put forward the design of the ramjet and obtained the patent, but at that time, there was no corresponding means of boosting nor proper materials, so the designs stayed on paper. Later, the V-1 cruise missile was launched by the Luftwaffe on December 24, 1942. Since then, the ramjet has been successfully applied to some missiles and aircraft in supersonic flight.

The type of ramjet that burns at a subsonic speed in the combustion chamber is only suitable for supersonic flight. For hypersonic flight, this kind of engine does not function properly. The main reason is that the airflow (stagnated from hypersonic to subsonic in velocity) results in air intake loss, which can send the engine's performance into sharp decline. Air stagnation can also lead to excessively high temperatures of gas, exceeding the loading limit of the combustion chamber. In addition, the dissociation of gases caused by high temperatures also consumes a large part of the chemical reaction's heat. When the Mach number of the aircraft exceeds 5, the air in the inlet slows down to raise the temperature in the engine, and the combustion can no longer deliver the energy. In fact, Mach 5 is the actual limit of the ramjet. In order to break through this limitation, the super-combustion ramjet engine emerged. The engine uses the oblique shock wave generated by the head and forebody of the aircraft to properly compress the incoming gas so as to reduce its speed and increase its temperature, but it still flows at supersonic speed after reaching the combustion chamber.

Scramjet fuel is composed of hydrogen and hydrocarbon. Compared with hydrogen fuel, hydrocarbon fuel does not react as easily, and contains less energy per unit mass. Moreover, when it is used to cool the thermal structure, its heat capacity is low and its cooling effect is poor. However, lower speed aircraft use hydrocarbon fuel, so its infrastructure has spread all over the world. Moreover, hydrocarbon fuel is easy to carry and provides a lot of energy per unit volume. Compared with hydrogen fuel with the same energy content, hydrocarbon fuel needs less onboard space. This is why hypersonic missiles usually use it. In a scramjet cooled by fuel, the fuel needs to be used as a heatsink (heat absorbing substance) to deal with the excess heat. In a heat balance system, the amount of fuel needed to absorb the excess heat of the structure should be less than the amount needed for combustion. When hydrocarbon fuel is used, it can reach equilibrium at Mach 8. From this point of view, if the Mach number is greater than 8, hydrogen is the best fuel.

4.3 Types of scramjet

After years of development, many kinds of scramjet solutions have been studied and designed abroad, including the pure scramjet, subsonic/supersonic combustion Dual Mode RamJet (DMRJ), subsonic/supersonic dual combustion chamber ramjet, dual fuel ramjet, injection scramjet, integrated rocket liquid fuel scramjet, and solid Dual Mode Ramjet (DMRJ). Among them, the foremost problem is how to combine the subsonic combustion engine and the supersonic combustion engine to smoothly convert between the two combustion modes while reducing weight. At present, three types have been proposed, namely the dual-mode ramjet (DMRJ), dual combustion chamber ramjet, and dual fuel ramjet.

(1) The subsonic/supersonic combustion DMRJ

The subsonic/supersonic combustion DMRJ is a kind of engine that can work in two modes: subsonic combustion and supersonic combustion. When the Mach number of the engine is lower than 5, positive shock waves are generated in the inlet of the engine, causing the air flow at the inlet of the combustion chamber to be subsonic and achieving subsonic combustion. When the Mach number is greater than 5, a series of oblique shock waves are generated in the inlet of the engine, and the airflow enters the combustion chamber at supersonic speed to achieve supersonic combustion. Thus, the lower limit of the Mach number of the DMRJ is reduced to 2.5–3, which expands the working range of the engine.

Taking advantage of a simple structure and light weight, the subsonic/supersonic combustion DMRJ can use either hydrogen fuel or hydrocarbon fuel. It is very complex in design, with the airflow geometry of the engine, the injection fuel position, and the injection amount set to achieve the conversion of two combustion modes. An isolator between the inlet and combustion chamber is an indispensable part of the dual-mode scramjet, acting as a pneumatic thermal buffer section between the two. It can isolate the interference between the inlet and the combustion chamber, and also make the engine transition smoothly from the subsonic combustion mode to the supersonic combustion mode. Whether in the subsonic or supersonic combustion state, the isolator can withstand the back pressure change of the downstream combustion chamber under different combustion states without affecting the flow state of the upstream inlet, so as to ensure that the inlet will start. Generally, a shock wave string will be formed in the isolator to provide the required air with a certain pressure, temperature, speed, and flow rate for the combustion chamber. The scramjet can be used in hypersonic missiles, aircraft, and aerospaceplanes.

Following the cancellation of the National Aerospaceplane (NASP) program, NASA developed an experimental uncrewed hypersonic aircraft, called the X-43A. It uses a hydrogen-fueled subsonic/supersonic combustion dual-mode ramjet. Although the flight test of the X-43A was successful, its flight Mach number was fixed at 7 or 10.

Developed by the Air Force Research Laboratory (AFRL) in collaboration with DARPA, the X-51A experimental aircraft is powered by a JP-7 hydrocarbon fuel subsonic/supersonic combustion two-mode ramjet engine. The X-51A has conducted four flight tests. During the fourth experimental flight on May 1, 2013, the flight reached Mach 4.8. The X-51A detached from the solid rocket and ignited its own subsonic/super-sonic combustion two-mode ramjet engine. Within 240 seconds, the fuel in the engine was exhausted and the test aircraft accelerated to maximum of Mach 5.1. The X-51A flew more than 426 kilometers during the experimental flight, and acquired flight data of 370 seconds.

(2) The subsonic/supersonic combustion dual combustor ramjet
For the scramjet that uses hydrocarbon fuel, when the engine works within Mach 3 to Mach 4.5, it is difficult to ignite the fuel. In order to solve this problem, the concept of the subsonic/supersonic combustion dual combustor ramjet was put forward. The inlet of the engine is divided into two parts: one guides the airflow into the subsonic combustor, and the other guides the rest of the airflow into the supersonic combustor. The subsonic combustor located at the front plays the role of an ignition source for the supersonic combustor, which can release the heat of the fuel at a low Mach number. From another point of view, the combustion of the gas flow in the scramjet combustor is supplementary, after subsonic combustion. There is no instability in the combustor inlet thanks to the rich oil mode of the subsonic combustor. This scheme is suitable for hypersonic missiles because of its small technical risk and low development cost. The US Navy's Hypersonic Flight Demonstration Program (HyFly) developed a dual combustor ramjet, but failed three consecutive flight tests.

(3) The subsonic/supersonic dual fuel ramjet
To overcome the long ignition delay of hydrocarbon fuel, hydrogen is used to ignite it. At low Mach numbers, hydrocarbon fuel is used as subsonic combustion, and at higher Mach numbers (such as 5 or 6), hydrogen is used as fuel. This kind of engine can be used in hypersonic aircraft and aerospaceplanes, but it is still in the laboratory stage due to its complex structure.

4.4 The technical obstacles of the scramjet

Although the scramjet avoids the problem of excessive airflow temperature suffered by the subsonic combustion ramjet, it also poses many difficulties. The technical barriers to the development of the scramjet are as follows:

1) Fuel mixing, ignition, and stable combustion
   Since air stays in the engine for only a few thousandths of a second, igniting it is harder than lighting a match in a tornado, let alone making sure it burns steadily in

such an environment. Therefore, to fully mix the airflow and fuel in the combustion chamber, successful ignition and stable combustion are the primary aims.

2) *The thrust of the engine*

Whether the engine thrust can be greater than the resistance of combustor plus the external drag of the aircraft becomes the key to the successful operation of the scramjet. Although the external drag of the aircraft can be greatly reduced if the integrated design of the wave-riding shape body and scramjet engine are adopted, the internal resistance of the combustion chamber will be increased due to the use of hydrocarbon fuel (its calorific combustion value is 2.8 times lower than that of hydrogen fuel). In addition, the possible methods of increasing mixing and ignition as well as increasing the isolator also raises the internal resistance of the combustion chamber. When the velocity increases, if the coefficient of the drag remains unchanged or decreases slightly, the resistance increases with the square of the velocity, while the thrust increases only with the first power of the velocity. Thus, thrust may be insufficient, or the thrust margin may be small.

3) *Transitioning from subsonic combustion to super combustion*

To deal with this problem, the schemes of the dual combustion chamber, dual fuel, and variable geometry have been suggested, but they will increase the complexity of technology and the weight of the engine. Thus far, the most studied is the subsonic combustion/super combustion dual-mode ramjet (DMRJ). Because its flow field is too complex and sensitive to external disturbance, it requires repeated tests to achieve full success.

4) *The regenerative cooling problem*

The combustor of the scramjet will be subjected to both aerodynamic heating and combustion heat release during operation, and the temperature of the combustor is very high – up to 2,700 K at Mach 6. At present, regenerative fuel cooling technology is used for heat protection in the combustion chamber. To this end, there are high requirements for fuel selection, heat transfer efficiency, and material technology.

5) *There is no suitable ground test equipment to properly simulate a scramjet engine, and there is no way to extrapolate ground test data to real flight conditions*

The key scramjet technologies include fuel injection, mixing and ignition, combustion chamber design and test technology, engine cooling, and engine body integration design.

4.5 The combined propulsion system

When an air-breathing engine is used in the space vehicle, its effective operating range must be Mach 0–25, and its flight altitude should be from sea level to the outer atmosphere. When such an engine is used in hypersonic missiles, its flight Mach number is 0–8. When an air-breathing engine is used in hypersonic aircraft, its flight Mach number will be 0–6. Based on the above demands, there are two concepts for the combined propulsion system currently under discussion around the world. The first is the Turbine-Based Combined Cycle (TBCC), in which the turbojet engine works in the range of Mach 0–3; the ramjet engine works in the range of Mach 3–5, while the scramjet engine works when Mach 5 or more. This makes the structure of this propulsion system relatively complex. Another concept is the Rocket-Based Combined Cycle (RBCC). The rocket system is adopted to replace the turbojet system of the TBCC propulsion system, which features simple structure and easy integration. However, it also lacks the advantages of the air-breathing engine. For aerospaceplanes requiring single-stage-to-orbit (SSTO), a rocket engine is also used when it accelerates to Mach 15. For two-stage-to-orbit (TSTO) vehicles, rocket engines will be used in the upper stage.

The USA has conducted a number of research projects on TBCC over a long period of time. In 2016, DARPA revealed plans to launch a project called Advanced Full Range Engine (AFRE) in 2017 to demonstrate the mode transition of a full-scale TBCC by an off-the-shelf turbine engine, to evaluate the feasibility of TBCC propulsion system engineering for hypersonic aircraft. On August 12, 2016, DARPA announced a Broad Agency Announcement (BBA) on the AFRE project, which explained the background, development goals, planning, and performance requirements of the project in more detail.

4.6 The deeply precooled SABRE engine

As space powers invested heavily in the development of scramjet engines, the Synergistic Air-Breathing Rocket Engine (SABRE) from Reaction Engines Limited (REL) has completed more than 100 bench tests. It can cool the air flow from 1,000°C to −150°C in less than one hundredth of a second without frost blockage, opening a new path for the development of hypersonic propulsion systems, and completely changing the game in its field.

The birth of the SABRE engine can be traced back to the mid-1980s. At that time, in order to reduce the cost of round-trip transport between the Earth and space, many countries came up with plans for horizontal take-off and landing in aerospaceplanes. Among them, the most renowned were NASP (using a scramjet engine) in the United States, SANGER II (using a turbo-ram jet engine in Germany), and HOTOL (using a pre-cooled jet engine) in the UK. The most important difference was their propulsion systems. By the 1990s, all three projects had been canceled, but the three main players who had participated in the development of HOTOL in the UK still insisted on pushing forward. In 1989, the Reaction Engines Limited (REL) was established for self-financing and low-key implementation of the SKYLON aerospaceplane program. A large number of technological innovations were carried out, and major progress was made in the development of SABRE engine.

The SABRE engine has a dual-mode capability. In rocket mode, the engine operates as a closed-cycle liquid oxygen/liquid hydrogen high specific impulse rocket engine. In air-breathing mode (from take-off to Mach 5), the liquid oxygen flow is replaced by atmospheric air, increasing the installed specific impulse three to six times. The airflow is drawn into the engine via a 2-shock axisymmetric intake, and is cooled to cryogenic temperatures prior to compression. The hydrogen fuel flow acts as a coolant for the closed-cycle helium loop before entering the main combustion chamber. Chilled helium is used to cool the air.

A key component of the SABRE engine is a pre-cooler heat exchange system that cools the air intake. The heat exchanger consists of thousands of thin-walled tubes through which coolant is passed. When airflow passes through the heat exchanger, it is cooled by helium flowing in the pipes. REL has mastered two key technologies for this purpose, one of which is related to manufacturing. The heat exchanger is made of Inconel 718 with precooler tubes of a 0. 88mm bore diameter and a tube wall thickness of 30 microns, which can ensure heat exchange performance without a reduction in physical strength. The second is the frost control technology, which stops it becoming blocked with ice. If the SABRE engine is successfully developed, the two-stage-to-orbit (TSTO) aerospaceplane can be developed on its basis, and can then evolve into the single-stage-to-orbit (SSTO) aerospaceplane Skylon.

Reaction Engines Limited (REL) has designed a Scimitar liquid hydrogen precooled engine originated from SABRE. The Scimitar engine can be used in the supersonic airliner of the EU's Long-Term Advanced Propulsion Concepts and Technologies (LAPCAT). LAPCAT aircraft can fly at 6,437 km/hr – about 2.5 times the top speed of Concorde. Passengers will be able to fly from London to New York in just two hours. In addition, LAPCAT aircraft can fly at the altitude of 28 kilometers. Of course, Scimitar engines can also be used in military aircraft.

Deep precooling offers many benefits to hypersonic engines. First of all, it reduces the inlet air temperature of the engine, increases the flight Mach number of the aircraft, and relieves the thermal environment of each working component. Secondly, reducing the temperature of the inlet airflow can increase the density of the gas and increase the flow rate, thus increasing the thrust. Compared with the aforementioned TBCC and RBCC combined engines, the advantages of the deeply precooled SABRE engine are as follows: wide range of flight speeds from Mach 0 to Mach 20; with a higher thrust-to-weight ratio, its specific impulse is higher than the combined engine in the whole range of Mach numbers, the thrust-to-weight ratio is 9–14 at Mach 2, and the thrust-weight ratio is 6 at Mach 5; and compared to the TBCC integrated engine, it can reduce fuel consumption by 18%–23% at all Mach numbers. With its single engine, it can replace multiple engines or even combined cycle engines, and can greatly reduce the total take-off weight. Compared with the subsonic and supersonic combustion dual-mode ramjet, it avoids very complicated flow and combustion controls, and reliability is enhanced. SABRE and Scimitar engines are powered by hydrogen fuel, and both engines deliver sufficient thrust, as the calorific combustion value of hydrogen is 2.8 times that of hydrocarbon fuel. However, the price of hydrogen is relatively high. Due to the phenomenon of hydrogen embrittlement, stronger materials are needed.

On March 25, 2019, British Reaction Engine Co., Ltd. (REL) successfully completed high-temperature testing of the SABRE engine's full-size precooler (HTX) prototype in Mach 3.3 flight at the TF2 test facility at the Colorado Air and Space Port, USA. Follow-on tests were planned for Mach 4.2 and Mach 5-simulated high-temperature airflow at the TF2 test facility in the months that followed.

REL claimed that it had raised 170 million US dollars, including 11.1 million in contracts with the European Space Agency (ESA), 66.2 million promised by NASA, 26.5 million invested by BAE systems, and about 66.2 million in private investment raised when the company was founded. The company apparently had enough money to support ground tests of the SABRE engine in sub-scale by 2020, and signed a partnership agreement with the US Air Force for SABRE engine technology, conducting research on its potential applications and development. As a result, REL is establishing a US branch in Colorado to strengthen its ties with US government departments and industry partners.

Deep precooling technology is highly appreciated around the world. It is regarded as a disruptive technology in the engine field, and the second revolution after the invention of jet technology. The UK's Universities and Science Department believes that the technology could revolutionize human travel in the air and space, while the US Air Force Research Laboratory (AFRL) considers the SABRE engine to be an attractive technology that is technically feasible and could be used in two-stage-to-orbit (TSTO) aerospace aircraft or national defense. According to the team at the Aircraft Engine Aerothermodynamics National Laboratory, led by Chen Maozhang from

Beijing University of Aeronautics and Astronautics, SABRE is a significant technical breakthrough that will change the appearance of the whole aircraft engine and hopefully become the most applicable hypersonic power technology of the future.

China has developed rapidly in hypersonic technology, but there are few start-ups in this area. In the long run, it must develop original and disruptive technologies and follow the path of innovative development.

References

[1] Huang Zhicheng. *Sky and Sky Vision* [M]. Beijing: Electronic Industry Press, 2015.

[2] Huang Zhicheng. *The Fourth Wave of Aerospace Science, Technology, and Society* [M]. Guangzhou: Guangdong Education Press, 2007.

[3] Huang Zhicheng. *The Development of a Round-the-world Transportation System: High-tech Status and Development Trends* [M]. Beijing: Science Press, 1993.

[4] Chen Haipeng, Zhang Bing, Yi Yi, et al. 'A Review of the Development of American Commercial Aerospace Travel' [J]. *China Aerospace*, 2013 (9): 32–38.

[5] Huang Zhicheng. 'The United States begins the commercial era of manned spaceflights' [J]. *Space Exploration*, 2014 (11): 17–21.

[6] Zhang Rui. 'The Commercialization of American Manned Spaceflights' [J]. Spacecraft Engineering, 2011, 20 (6): 86–93.

[7] An Hui. 'Analysis of the development of the US military aerospace shuttle' [J]. *Satellite Applications*, 2010 (5): 49–54.

[8] Huang Zhicheng. "Dream Catcher': a new generation of US space shuttles' [N]. *Wenhui Bao*, 2016-01-24 (7). Xinhua Digest, 2016 (8): 124–126.

[9] Huang Zhicheng. 'Hypersonic technology: the pursuit of humanity in the new century' [J]. *Space Exploration*, 2003 (6): 22–25.

[10] Cai Guozhen, Xu Dajun. *Hypersonic Vehicle Technology* [M]. Beijing: Science Press, 2012.

[11] Wang Zhenguo, Liang Jianhan, Ding Meng, et al. 'The progress of power system of hypersonic aircraft' [J]. *Advances in Mechanics*, 2009, 39 (6): 716–739.

[12] Huang Zhicheng. 'The mystery of ram engines' [J]. *Aviation Knowledge*, 2007 (10): 30-31.

[13] Huang Zhicheng. "Sabre': A New Motive for Sky and Earth Travel [J]. *International Space*, 2015 (7): 11–12.

[14] Huang Zhicheng. 'The military value of hypersonic technology' [J]. *Space Exploration*, 2017 (4): 26–29.

[15] Zou Zhengping, Liu Huoxing, Tang Hailong, et al. 'Research on Strong Precooling Technology of the Hypersonic Aero Engine' [J]. *Chinese Journal of Aeronautics*, 2015, 36 (8): 2544–2562.

[16] Dora Musielak. *Hypersonic Spy planes, Civil Transports, and Spaceplanes: Projecting the Future of Transcontinental Flight and Access to Space* [R/OL]. https://info.aiaa.org/tac/PEG/HSABPTC/Public.

[17] Federal Aviation Administration. The Annual Compendium of Commercial Space Transportation: 2017 [R/OL]. https://brycetech.coMa/downloads/FAA_Annual_Compendium_2017.pdf.2017

[18] NASA. Commercial Crew Program Overview [R/OL]. Masters Forum 20 Maria Collura. https://www.nasa.gov/pdf/552848Maain_Commercial_Crew_Programme_Overview_Collura.pdf.2011

[19] Keith Reiley, Michael Burghardt, Jay Ingham, Michael F. Lembeck. *Boeing CST-100 Commercial Crew Transportation System* [R]. AIAA 2010-8841.

[20] Arthur C. Grantz. *X-37B Orbital Test Vehicle and Derivatives* [R]. AIAA 2011–7315.

CHAPTER 9

THE OPPORTUNITIES AND CHALLENGES OF COMMERCIAL SPACE TRAVEL IN CHINA

In the space of half a century, China's aerospace industry has grown from nothing to a position of strength. Its remarkable achievements have attracted worldwide attention. At present, with the tide of development of global commercial aerospace, it faces both opportunities and challenges. To deal with this new situation, China will deepen system reform based on foreign examples of commercial aerospace development. State-owned enterprises, private companies, and hybrid firms will join the innovation-driven civil-military integration of commercial aerospace, creating a future that is full of potential.

1. Lessons Learned from the Development of Commercial Space Travel in the United States

The development of commercial aerospace in the United States has significantly enhanced the vitality and competitiveness of the sector, and has achieved preliminary results. American commercial aerospace has the ability to design, manufacture, launch, and operate aerospace products to meet the needs of the government, military, corporations, scientific research institutions, and other users to purchase services from the market. The implementation of commercial aerospace has added new driving forces to the future development of American aerospace overall.

First of all, the development of commercial aerospace has reduced the cost of space travel, as its business model significantly reduces costs and improves cost performance. SpaceX announced the launch prices of its Falcon 9 and Falcon Heavy rockets on May 2, 2016. The list price of launching a Falcon 9 rocket stood at 54 million US dollars, about one third of the launch price of similar launch vehicles in the United States. The list price of launching a Falcon Heavy rocket was about 90 million US dollars, far below the price of 350 million US dollars to launch a same-class heavy Delta IV rocket. In terms of R&D costs, NASA released a report stating that if it develops a Falcon 9 rocket, it will cost at least 1.3 billion US dollars, while if SpaceX develops one, it will cost less than 400 million. With the successful reuse of the Falcon 9 rocket, its launch cost will be further reduced. SpaceX has found favor in the aerospace market with its low costs, and its Falcon 9 rocket has won 38 launch contracts worldwide for the next five years, including 14 from the government and 24 from the international commercial launch market.

NASA has drastically reduced the cost of shuttling back and forth to the Space Station through commercial orbit transport services and other projects. The 12 commercial cargo resupply flights to the ISS by Dragon spacecraft were valued at about 1.6 billion US dollars, with an average cost of 133 million per flight – far below than the average cost of 1.5 billion US dollars per flight of the space shuttle. While the Russian Soyuz costs 70 million US dollars per seat, SpaceX is developing a manned Dragon 2 spacecraft for an estimated 20 million per seat.

With similar reconnaissance capabilities, the cost of commercial remote-sensing satellites is only around a fifth less than that of military satellites, and the service life can be prolonged. The United States spends only 380 million US dollars each year to obtain high-resolution commercial remote-sensing data through its enhanced observations program.

Second, the development of commercial aerospace improves the efficiency of aerospace activities. In order to obtain competitive advantage and more profit, commercial aerospace must pay attention to improving efficiency. Compared with traditional aerospace, commercial field is more efficient in organization management, product development, and technological innovation. Commercial aerospace companies simplify their management organization and streamline work procedures by establishing a flat organizational structure, making the connection between R&D and production closer. It has been proved that flat management improves the efficiency of the whole R&D team and saves management costs. SpaceX has long maintained the scale of hundreds of employees, and has independently developed the Falcon 1 and Falcon 9 rockets as well as the Dragon spacecraft.

By shortening the development cycle, commercial aerospace companies can launch products rapidly in order to occupy the market. SpaceX's Falcon 9 rocket took only four and a half years from its design to its first flight. The Dragon spacecraft's development began in 2005, and the first demonstration flight test was conducted in 2010. In 2012, the company completed two docking flights between the Dragon spacecraft and the ISS, with an interval of less than five months.

By the end of July 2017, the Falcon 9 rocket had been launched 38 times, included 11 times in 2017, and three times for reuse of the first stage. The first-stage recovery was successful 13 times, and recovery on board was successful eight times.

Third, the development of commercial aerospace has accelerated the pace of technological innovation. The technological innovation boom of commercial aerospace companies continues to grow, and a number of disruptive technologies have made new progress, which will profoundly affect and change the development trend of space technology in the future. SpaceX's breakthrough in the recovery of its Falcon 9 rocket is an example of disruptive innovation in space technology. Looking to the future, the vitality of technological innovation in commercial space travel will push space technology to break bottlenecks and enable space exploration to go faster, further, and better.

Finally, the development of commercial aerospace has promoted social and economic development. Commercial space travel has played an important role in promoting the development of the national economy. Operating with a market model and a large number of participants, commercial aerospace promotes employment while driving the development of related upstream and downstream industries.

The lessons learned by the United States in the development of commercial aerospace are as follows.

1.1   Driven by two engines: the government and the market

The United States leads the world in commercial space travel. The US government has implemented a national policy to encourage and support the development of commercial space flight. Published recently, the National Space Policy and the National Space Transportation Policy have proposed to develop commercial aerospace, promote the innovation and international competitiveness of America's space industry, and encourage the purchase of commercial aerospace products and services to the maximum extent to meet the needs of government and society.

Perfect legislation provides a legal guarantee for commercial aerospace. The US government has developed relatively comprehensive laws and regulations in various fields of commercial aerospace to allow companies in the sector to have legal compliance. The United States has clear legal provisions in fields such as commercial launch, satellite navigation, communications, and remote-sensing. The entry conditions, codes of conduct, legal responsibilities, and rights of market subjects are clarified through legal provisions. The policies, laws, and regulations are consistently improved in the process of implementation in order to promote an upgrade in capabilities within American commercial aerospace. For example, in the field of commercial remote-sensing, the United States has revised its commercial remote-sensing policies and laws four times, allowing commercial companies to increase the resolution of optical satellite images sold to the market from 1 m to 0.5 m, and now to 0.25 m. The resolution of radar remote-sensing satellite images has increased from 3 m to 1 m, significantly enhancing the competitiveness of American commercial remote-sensing satellites in the global market.

In October 2015, the National Geospatial Intelligence Agency (NGA) released the 2015 Commercial GEOINT Strategy, seeking to leverage the technological advantages of the commercial satellite industry, operators, and data analysis companies to improve procurement flexibility. In November 2015, then President Barack Obama signed the US Commercial Space Launch Competitiveness Act, allowing US commercial companies to bring mineral resources from small planets and other extraterrestrial objects back to the Earth, which further stimulated commercial companies' interest in deep space exploration. In January 2016, the National Oceans and Atmospheric Administration (NOAA) issued the first Commercial Aerospace Policy, stating the background, objectives, principles, measures, and regulatory agencies to show how commercial aerospace ability can be used for meteorological observation. This policy has taken an important step towards breaking the long tradition of relying solely on the government to provide meteorological services in the United States.

In addition, giving full play to the role of the market is the key to the development of commercial aerospace. NASA has chosen several commercial aerospace companies to compete in various projects, including commercial orbital payload and crew transportation. In this fierce competition, SpaceX stands out for its innovation ability.

At the same time, the prosperous capital market provides strong financial support for the development of commercial aerospace. Commercial aerospace projects are often funded by Silicon Valley venture capital in the early stage, with strategic investors introduced in the middle stage, and financing from the capital market in the later stage. This has been a huge boost to the success of commercial space travel. For example, DigitalGlobe's WorldView-2 satellite has been successfully developed and launched through private equity.

1.2 Encouraging entrepreneurship and innovation

Future space missions are full of challenges and risks, and only by strengthening disruptive technological innovations will it be possible to find a low-cost and low-risk way to space. The development of commercial space travel has opened the aerospace industry to the public to brainstorm ideas and accelerate the innovation of disruptive space technologies. At the heart of commercial space flight is innovation. Compared with traditional space flight, the advantage of the new mode is the fact that it gives full play to the innovations of entrepreneurs.

The success of SpaceX and Blue Origin proves that commercial aerospace development also requires entrepreneurial ambition, enthusiasm for exploration, persistence, and the ability to defy setbacks. The creation and development of SpaceX originated from Elon Musk's vision of colonizing Mars. Blue Origin was founded and grew out of Bezos' desire to send heavy industry detritus from chemical works, cement plants, and steel plants into space to eliminate pollution and emissions. China also needs entrepreneurs with visions and trailblazing courage. In this regard, it should first encourage entrepreneurs with relevant ideas and pioneering attitudes to work towards making their dreams into reality. China should also guide entrepreneurs and investors to treat success or failure rationally, and advocate concentration. It is inevitable that an entrepreneur debuting in the high-risk commercial aerospace will suffer setbacks. Regardless, entrepreneurs and investors should see setbacks not as failures but as learning opportunities.

1.3 In-depth development of civil-military integration

Many commercial aerospace projects in the United States (such as high-resolution remote-sensing satellites) are closely related to national security and the military. However, the USA has not stopped making progress, but instead has made laws and regulations and enhanced supervision to make these commercial aerospace projects into a model of in-depth civil-military integration.

Since 2000, the Iridium system has been used extensively by the US military to conduct battlefield communications and support operations, with information provided by commercial remote-sensing satellites in several local wars. During the conflict in Afghanistan, the price of commercial remote-sensing satellites purchased by the United States (such as IKONOS) was only 20 US dollars per square kilometer. Commercial remote-sensing satellites meet military needs, and also significantly reduce the cost of military missions and expand their applications in the national economy.

SpaceX finally certified the qualification of national security space launch missions in May 2015, and successfully launched the first military reconnaissance satellite with the Falcon 9 rocket in May 2017. In doing so, it broke the long-term monopoly of the United Launch Alliance (ULA) on the US government payload launch service, and became the first commercial aerospace company to enter the US military aerospace launch market. On September 7, 2017, SpaceX's Falcon 9 rocket launched US Air Force's X-37B small spaceplane into orbit.

1.4 Leveraging the role of the National Space Center

Being responsible for the development of public space travel, NASA has played an irreplaceable role in the development of commercial space travel in the United States. First of all, it has injected start-up funding into commercial aerospace companies through a number of commercial aerospace programs, and supports the operation of commercial aerospace companies with prepayment of service contracts.

Secondly, in order to improve innovation in disruptive technology, NASA has carried out institutional innovation internally, and established a new Space Technology Mission Directorate (STMD) whose mission is to invest widely in society to develop bold and broadly applicable disruptive technology that the industry currently cannot achieve.

Thirdly, NASA has accelerated the transformation of its technologies by developing commercial aerospace. For example, many key technologies of the Merlin rocket engine used by SpaceX's Falcon rocket are derived from the kerosene/liquid oxygen (LOX) fuel rocket engine of the lunar orbiter developed by NASA in the 1960s. After a series of improvements, the latest version of the Merlin engine has achieved high performance. Bigelow Aerospace's capsule with expandable habitat technology has also been one of NASA's technologies for many years. Sierra Nevada's small Dream Chaser spaceplane uses the HL-20 lifting body configuration developed by NASA based on years of accumulated lifting body technology.

Finally, through its aerospace education activities, NASA has improved the public's literacy in terms of aerospace science and technology, and has trained a large number of innovative personnel for the development of commercial aerospace.

1.5 Improving safety

In the process of developing commercial space travel, many start-ups got off to a difficult start due to major accidents caused by design defects and schedule issues. Although these accidents had a significant impact on the development of the companies, their founders were not discouraged. Instead, they learned from their mistakes and paid more attention to the strict requirements for reliability and safety in the aerospace industry.

For example, on October 28, 2014, the Antares carrier rocket (launched by the US Orbital Sciences Corporation) exploded six seconds after ignition and lift-off, and destroyed the rocket and satellite at the same time. On October 31, 2014, Virgin Galactic's SpaceShipTwo (SS2), carried by White Knight 2, took off from the Mojave Air and Space Port. During the flight, the co-pilot died due to an accident and the pilot was seriously injured. On June 28, 2015, SpaceX carried out the launch mission of the Dragon spacecraft. Due to a fault in the steel strut that was holding down a helium bottle, the rocket exploded and disintegrated 139 seconds after it launched. On September 1, 2016,

SpaceX's Falcon 9 rocket, which was scheduled to be launched two days later and was loaded with the Amos-6 communications satellite, exploded at Launch Complex 40 (LC-40) at Cape Canaveral Air Force Station, destroying the rocket, satellite, and launchpad.

The impact of these accidents is far less than that of the two major space shuttle crashes in the traditional space age in the United States. The economic were covered by insurance companies, and there was no political impact. Nor did these incidents halt the advance of American commercial space travel. Rather, they provided a profound lesson that both innovation and safety should be emphasized in development, requiring a balance between cost, schedule, and quality control.

2. The Opportunities and Challenges of Commercial Space Travel in China

In the transition period of global aerospace development, due to the issues caused by the Soviet Union and Western countries during the transition period, and the aftereffects of the Cold War, China's aerospace development is facing a period of strategic opportunity. With the rapid development of its economy, it has made a series of achievements in manned space, including the BeiDou Navigation System, the high-resolution remote-sensing satellite, and lunar exploration. It has shortened its gap with advanced global standards, and is developing confidence in its own independent development of space technology.

At present, the status of China's economic development means that its aerospace industry is facing difficulties in slowing the growth of funds and increasing costs. With the comprehensive commercialization of space in the United States and around the world, China's aerospace development will face severe challenges and new tests.

2.1 The opportunities of commercial aerospace in China

The development trend of global commercial aerospace is currently spreading from the United States and Europe to China. State-owned, private, and diversified ownership aerospace enterprises are stepping up the development of commercial aerospace, initially forming a competitive mechanism. The State Council made a decision to encourage private capital to enter the space field, which is sure to promote in-depth reform of China's space system and lead to a new boom in commercial aerospace development.

In recent years, China's national and local governments have issued a series of supporting policies in the application of aerospace commercialization. The National Development and Reform Commission, the Ministry of Finance, the State Administration of Science, Technology, and Industry for National Defense, and relevant departments jointly issued the 'Medium- and Long-term Development Plan for China's Civil Space Infrastructure (2015–2025)' in 2015, which proposed 'supporting private capital to invest in satellite development and system construction', 'combining national and social investment for public welfare and commercial programs, and commercial projects based on social investment', and 'encouraging and supporting qualified enterprises to invest in the construction of satellites within the planning plan', providing policy guidance for the development of commercial aerospace. Special support policies have also been introduced in satellite navigation applications and remote-sensing satellite applications, such as the 'Medium- and Long-term Development Plan for China's Satellite Navigation Industry', the 'Notice on Organizing and Conducting the Major Application Demonstration Development Project of the BeiDou Navigation Satellite', 'Some Opinions of the National Administration of Surveying, Mapping, and Geoinformation on the Promotion and Application of the BeiDou Satellite Navigation System', and 'Some Opinions of the State Council on Promoting Emerging Consumption and Expending Domestic Demand'.

On November 27, 2014, the State Council officially released 'Document No. 60', which encouraged private capital to enter the aerospace industry and participate in the construction of national civil space infrastructure. This improved the data policy for civil remote-sensing satellites and government procurement services, encouraged private capital to develop, launch, and operate commercial remote-sensing satellites, and provided market-oriented and specialized services. It guided private capital to participate in the construction of a satellite navigation ground application system. The State Council made a timely decision to encourage private capital to enter the space field, which promoted the in-depth reform of China's space system and led the tide of entrepreneurship in the nation's aerospace industry.

China's commercial aerospace development has now been backed by a strong industrial base. Since 1956, the nation's aerospace industry started from scratch, developing quickly. It rapidly increased in size and went through phases of construction, adjustment, reform, and development, building an effective aerospace system engineering management system and operation mechanism. It also established a systematic system supporting scientific research and production innovation, and a high-quality professional talent team. With the state-owned CASC (China Aerospace Science and Technology Corporation) as its main body, China's aerospace industry has built a complete industrial chain on satellite manufacturing, satellite launch systems, satellite operation services, and ground equipment manufacturing and services. In this way, it has formed an industrial system with a complete range of categories and products, large overall scale, and strong comprehensive strength. It has made outstanding contributions to China's successful development of launch vehicles, satellites, manned space travel, and deep space exploration.

With the consistent progress of China's aerospace technology and the rapid development of its economy and society, various sectors have put forward more extensive demands for the commercial application of aerospace. For example, in the application of satellite navigation, more reliable, accurate, and cross-fusion integration of innovative services are needed. In the aspect of remote-sensing satellite application, more precise, diversified, and time-efficient observation requirements are proposed. As for the application of satellite telecom and broadcast, the most urgent needs are larger capacity, wider coverage, and higher security. In the launch field, due to the development of satellites (especially small satellite constellations), the number of launches will be multiplied, and the launch cost will be greatly reduced. With the further implementation of the Belt and Road Initiative, the demand for commercial satellite launches along the route has also brought opportunities for the development of commercial aerospace in China.

2.2 The current situation of commercial aerospace in China

Since the beginning of commercial exploration in the 1990s, the China Aerospace Science and Technology Corporation (CASC) has successfully entered the international market, signing international orders for the export of 11 satellites with nine countries. Six satellite export projects (including the Nigerian Communication Satellite and the Venezuelan Remote-Sensing Satellite) have been successfully launched, and five satellites (including the Belarusian Communication Satellite, which is still being developed) are also about to be launched. At the same time, CASC has carried out a series of value-added services related to the export of ground equipment, the coordination and support of frequency and orbital positioning, and operation and technical training, in combination with the export of satellites

CASC has provided 51 international commercial launches with a total of 59 satellites for customers from 22 countries, including seven satellite exports, 40 launch services, and 12 carrying launches. At present, CASC has a total of seven Long March (LM) launch vehicles that can be used for commercial and carrying launch services, including both conventional launch vehicles and LM-6 and LM-11 launch vehicles with rapid mobile launch capability. These vehicles can achieve multiple services such as single-satellite, multi-satellites, and carrying launch.

In the field of satellite operation, China Satellite Communications Co. Ltd (China Satcom), a subsidiary of CASC, is the only autonomous and controllable communications satellite operator in China. Its revenue ranks sixth among global satellite operators, and it has 10 satellites in orbit. In the field of the satellite-based Internet, CASC plans to build a Hongyan constellation communication system consisting of 60 satellites in 2020. At the same time, the China Resources Satellite Application Center (CRESDA) at CASC has initially established a remote-sensing application system, and remote-sensing applications have entered business operation.

In December 2016, China's Long March-2D carrier rocket successfully launched a pair of high-resolution commercial remote-sensing SuperView-1 01/02 satellites. The satellites will be operated commercially by China Siwei Surveying and Mapping Technology Co., Ltd., which is affiliated to CASC. The planned Siwei SuperView commercial remote-sensing satellite system is a combined remote-sensing constellation composed of 16 optical satellites at 0.5-meter resolution, four high-end optical satellites, four radar satellites, and several video and hyperspectral micro-satellites. After the full constellation network is completed, hour-level high-resolution remote-sensing data can be obtained. The SuperView-1 satellite has a 0.5-meter panchromatic resolution, a multispectral resolution of 2.0 meters, an orbit attitude of 500 km, and a width of 12 km. It has a variety of imaging modes such as consistent strip, multiple strip splicing, multi-target, and stereo, and its image quality has reached the world-class level.

The China Aerospace Science and Industry Corporation (CASIC) originally focused on weapon development, but has made a major push into the commercial aerospace market since 2015. Located at the Wuhan National Aerospace Industry Base, CASIC's plan shows that it will continue to promote five major commercial aerospace projects, including Feiyun, Caiyun, Xingyun, Hongyun, and Tengyun. The time span of the five projects will exceed 10 years, and the total investment is expected to exceed 100 billion yuan, involving rockets, satellites, and spaceplanes. The Hongyun project is a satellite-based Internet constellation consisting of 156 satellites, providing broadband Internet access services around the world. The Tengyun project plans to develop a two-stage airspace aircraft.

On January 9, 2017 at 12:11 am, the general-purpose Solid Launch Vehicle KZ-1A developed by CASIC successfully launched the Jilin-1 smart video satellite 03, and carried two CubeSats to realize a three-satellite launch.

Many local governments in China are also focusing on the development of commercial aerospace. With the support of the Beijing municipal government, Twenty First Century Aerospace Technology Co., Ltd. has completed the launch of the Beijing-1 and Beijing-2 satellites. By the end of 2015, the Jilin-1 satellites were successfully launched, with a 0.7-meter resolution and a width of 11.2 km. It also carried a smart experimental imaging satellite and two video satellites into orbit.

Many private enterprises in China are also involved in commercial aerospace field, starting from the launch of small satellites. On October 27, 2014, Beijing Xinwei Technology Group Co., LTD. cooperated with Tsinghua University to launch the 'Smart' communication experimental satellite, and prepared to develop a constellation composed of 32 satellites. June 15, 2017 saw the successful launch of the first two video satellites (OVS-1A and OVS-1B) of the Zhuhai-1 remote-sensing small satellite constellation developed by Zhuhai Orbita Aerospace Control Engineering Co., Ltd.

In terms of commercial rockets, China has several start-ups such as the Beijing One Space Technology Group Co., Ltd. (One Space) and the Land Space Technology Corporation Ltd. (Land Space). Both are developing rockets capable of launching small satellites.

2.3 The challenges of commercial aerospace in China

Unlike the US commercial aerospace industry that is run by private enterprises, the state-owned CASC and CASIC are still the major force in the development of the commercial aerospace industry in China. Other state-owned military and civil enterprises, mixed-owned enterprises, and private enterprises are also involved. China must explore in-depth how to enhance cooperation between state-owned enterprises and private enterprises, how to carry out the mixed system reform of state-owned enterprises, and how to support space entrepreneurship from aspects such as technology, funds, and infrastructure. It is widely believed that entrepreneurship will promote the development of China's space industry, concentrating on major events with innovative suggestions from the public.

Aerospace and other military enterprises face challenges in the development of commercial aerospace. First of all, aerospace research institutes in China are the main force of innovation, and are also an important force in promoting cutting-edge aerospace technology outside of the laboratory and realize engineering application. Within the current organizational system, some scientific research institutes that were originally positioned to engage in aerospace sci-tech research are now managed by aerospace enterprise giants and have become subordinate units. This mechanism objectively weakens the overall planning, leading, and regulating ability of the government departments in charge of aerospace sci-tech business. As a result, relevant institutes lose their socially-oriented service functions and find it difficult to drive and support the innovation ability of commercial aerospace.

Secondly, for historical reasons, in the current system of science, technology, and industry for national defense, the research and manufacturing organizations of military enterprise groups are based on a vertical platform structure and industrial management mode divided by different products and equipment systems such as aviation, aerospace, ships, armored vehicles, and weapons. On the one hand, there is a certain degree of intersection between the scientific research and the processing technology of many products among the military industry groups, such as aviation, aerospace, and other industries involving the same, similar, or interlinked advanced functional materials, high-end standard parts, and special sensors. On the other hand, the collaborative innovation and open social integration of the military industry still lag behind, and many obstacles remain for other sci-tech defense industries to enter the aerospace field.

Finally, the transformation of sci-tech achievements in the aerospace sector involves a series of institutional arrangements and multiple processes, which is a complex engineering system related to national security and confidentiality. For a long time, the state has invested a lot of human, financial, and material resources, and has created many achievements in aerospace science and technology, but the transformation is not smooth and the conversion rate is low.

There have been nearly 100 aerospace start-ups in China working on satellites, constellations, applications, rockets, TT&C (Tracking, Telemetry and Control), and fundamental aerospace technologies. They have made many achievements such as the Jilin-1 remote-sensing satellite, the LM-11 rocket, and the KZ-1 rocket. Although they face difficulties in terms of insufficient capital and technical strength, what is more important is the self-construction of their entrepreneurial teams. Since aerospace is a high-tech, high-risk enterprise, China needs innovative and cooperative personnel to join the wave of entrepreneurship. At present, the challenges faced by start-up private aerospace companies are as follows:

1) Due to the lack of influential leaders in China's aerospace industry, the investment community is not willing to get involved.
2) Most private aerospace companies do not have enough funds to attract more staff to join. As a result, they are beset by issues regarding both funds and people.
3) The existing launch sites and launchpads belong to the military. In recent years, the tasks undertaken by the industry are increasing steadily. The lack of commercial launch facilities will limit the development of commercial aerospace.
4) There are no laws and regulations for satellite TT&C (Tracking, Telemetry and Control), so the management of commercial satellites has not been fully implemented.
5) The lack of strategic resource reserves for commercial aerospace, especially frequency resources, has become the biggest weakness in China's attempt to develop a global communication constellation.

6) The innovative ability of private commercial aerospace enterprises is insufficient, and some are still in the accumulation stage of basic technology.

3. Suggestions for the Development of Commercial Space Travel in China

Regarding the development of the industry around the world, commercial aerospace has become a leader in the new era. Therefore, China should formulate a development strategy as soon as possible on the basis of summarizing examples from abroad. In order to speed up the development of China's commercial aerospace, the following preliminary suggestions are put forward:

1) *Accelerate the improvement of laws and regulations for commercial aerospace*
   The example of foreign commercial aerospace development has proved that policy support plays an important role. Legislation for the aerospace industry in China is deficient. To realize the rapid development of commercial aerospace, policies and regulations must be urgently improved. First, China must consider the demands of the industry during the preparation of its Space Law. As a law designed to regulate domestic space activities, this legislation should act as a normative guide, clarifying the boundaries between the government and the market, different regulatory departments, and other rights and responsibilities, and ensuring the orderly development of commercial aerospace activities. Second, based on the actual situation, priority should be given to solving problems that have no legal basis in key areas. For example, in the field of international commercial launch, due to the lack of clear guidance within China's domestic law, when international commercial launch contracts encounter problems, they can only look to foreign laws for reference, which is not conducive to participating in international competition. At the same time, domestic social capital in the field of commercial aerospace launch will also increase doubts. Therefore, China needs to introduce policies for commercial aerospace launches as soon as possible. Third, according to the needs of development, the government should introduce some transitional promotion policies. In order to encourage civil forces to participate in commercial aerospace, the government should issue corresponding guidance or interim measures according to business development.

2) *Encourage cooperation between state-owned enterprises and private companies*
   As the backbone of China's aerospace industry, state-owned military industrial

groups (especially CASC and CASIC) must take the initiative to make reforms, break the self-enclosed and self-contained system, and cooperate with private enterprises to jointly seek new profit growth points in different fields such as satellite development, operation, data analysis, and commercial launch.

It is impossible for commercial space travel to achieve leapfrog development only by the participation of state-owned military industry groups. Some private enterprises are required to participate and find their respective positions in cooperation and competition to achieve orderly development. The development of China's private space industry should be based on China's reality rather than copying the American model. Although China's state-owned aerospace enterprises and private companies have their own advantages, the latter have only just begun, and their overall strength cannot be compared with that of state-owned enterprises. The two models should learn from each other and integrate. Therefore, China's commercial aerospace industry should follow the development mode of mixed ownership. There needs to be in-depth exploration of how to enhance cooperation between state-owned enterprises and private companies, how to carry out mixed system reform within state-owned enterprises, and how to support space entrepreneurship from aspects such as technology, funds, and infrastructure. China's space entrepreneurship will promote the development of its aerospace industry, concentrating on major events alongside innovations from the public.

For this purpose, commercial satellites should be included in the scope of national government procurement so as to form a beneficial complement to the satellite system invested by the state, to create conditions for emerging space enterprises to carry out satellite development and commercial launch, and to provide support in terms of personnel, equipment, and facilities. Based on their own advantages, private space enterprises should establish the concept of collaborative development and achieve sound development through cooperation with universities, research institutes, and military enterprises. Before entering the commercial aerospace industry, private enterprises should first understand the demands of the market and users, so as to identify their positioning in the aerospace industry chain. They should then seek complementary advantageous cooperative units according to development needs, and jointly explore and establish an executable, profitable, and sustainable business model.

3) *Encourage social capital investment*

Compared with other industries, investment in the aerospace sector features a long cycle and large capital demand. Attracting the participation of social capital is an important way of promoting the development of commercial aerospace around the world. First, a national commercial aerospace industry investment fund should be established, like the national investment fund for the integrated circuit industry. It should be jointly initiated by the government and the two major military industrial groups, and should attract participation from investment institutions and social capital, so as to jointly promote the development of the commercial aerospace industry and share the dividends brought by the development. Second, social capital should play a guiding role in commercial aerospace development. In order to ensure strong decision-making in terms of investment, an independent risk assessment method is required. The participation of social capital will illuminate the commercial aerospace industry chain, clarify the actual demands, find the most profitable direction, and guide the development of commercial aerospace.

4) *Build a national aerospace research center*
   In order to develop commercial aerospace and realize disruptive innovation of relevant technology, the dual engines of the government and market are required. Aerospace research institutes are now all affiliated to state-owned space enterprises, so it is impossible for them to provide sufficient technical support to private aerospace companies as competitors. Therefore, a national key laboratory should be set up in aerospace research institutes under the dual leadership of the National Aerospace Administration and private companies. A national aerospace research center can be set up when conditions permit.

5) *Develop aerospace infrastructure in the framework of civil-military integration*
   The military and governments should coordinate the construction of commercial launching sites or new launchpads under the framework of civil-military integration. The operation and maintenance of commercial launch facilities should adopt the market model. The military and governments should also jointly formulate regulations for TT&C (Tracking, Telemetry and Control), and encourage and guide private capital to enter the field of commercial satellite TT&C in a legal manner. Large-size ground test equipment (such as engine test benches, wind tunnels, environmental test equipment spacecraft, and large-size vibration tables built by CASC) should be open to commercial aerospace companies for paid services.

6) *Cultivate innovative and cooperative entrepreneurial teams*
   Compared with entrepreneurial aerospace teams in the USA, China is still in its infancy. It is relatively inexperienced in business, and also insufficient in innovation

ability. For this reason, China's private aerospace entrepreneurs should be prepared for setbacks and failures. Not all aerospace entrepreneurs can achieve success. Only a team that is highly innovative and cooperative will prevail. China's private aerospace enterprises should strike a balance between capital requirements and the laws of aerospace development, combining short-term goals with long-term planning. They should focus not only on the segmentation of the market and the identification of profit models, but should also promote practical technological innovation.

4.   A New National System for China's Manned Lunar Exploration

In early 2019, Chang'e-IV successfully landed on the far side of the Moon for scientific exploration, marking a major breakthrough in China's lunar exploration program and attracting worldwide attention. Today, the Moon has become an important goal for global aerospace development. Chinese experts have repeatedly expressed that in the next decade or so, a China-led multinational lunar research station will be set up at the Moon's South Pole. China will finally set foot on the Moon, cementing its position as a world power.

The United States' manned moon landing experienced a tortuous development path. Although it put a man on the Moon 50 years ago, its development has since stagnated. George W. Bush proposed the idea of returning to the Moon, but when Barack Obama took office, he set his sights on Mars instead. On December 11, 2017, Donald Trump signed the Space Policy Directive-1, instructing the United States to return to the Moon.

References

[1]   Huang Zhicheng. *Sky and Sky Vision* [M]. Beijing: Electronic Industry Press, 2015.

[2]   Huang Zhicheng. 'How Commercial Aerospace is Leading the Aerospace 2.0 Age' [J]. *International Space*, 2017 (3): 2–6.

[3]   Huang Zhicheng. 'Commercial space travel: a new driving force for space exploration in the United States' [J]. *International Space*, 2016 (5): 64–68.

[4]   Zhang Zhenhua, Bai Mingsheng, Shi Yong, et al. 'The Development of Foreign Commercial Aerospace' [J]. *China Aerospace*, 2015 (11): 31–39.

[5]   Wu Qin, Zhang Mengtian. 'Analysis of the development of US commercial aerospace' [J]. *International Space*, 2016 (5): 6–11.

[6]   Fan Chen. 'Analysis of the development trend of business models of foreign aerospace companies' [J]. *Satellite Applications*, 2015 (11): 41–43.

[7] Chen Haipeng, Zhang Bing, Yi Yi, et al. 'A Review of the Development of American Commercial Aerospace Transportation' [J]. *China Aerospace*, 2013 (9): 32–38.

[8] Zhang Rui. 'The Commercialization of American Manned Spaceflights (Part I)' [J]. International Space, 2016 (4): 62–68.

[9] Zhang Rui. 'The Commercialization of American Manned Spaceflights (Part 2)' [J]. International Space, 2016 (6): 44–48.

[10] Liu Yufei. 'Mixed reform: an open window for commercial aerospace and the commercialization of space' [J]. *Satellite and Network*, 2017 (1): 12–15.

[11] Liu Yufei. 'Concerns about the Commercialization of Aerospace in China' [J]. *Satellite and Network*, 2015 (8): 26–28.

[12] Zhang Zhenjun. 'Legal Thoughts on China's Commercial Aerospace' [J]. *China Aerospace*, 2015 (12): 8–12.

[13] Xu Aiguo, Zhang Guoting. 'A preliminary discussion on issues related to the measurement and control management of China's commercial aerospace [J]. *Journal of Aircraft Measurement and Control*, 2017 (6)